Key Stage 4 Revision

OnCourse

AQA

ASSESSMENT and
QUALIFICATIONS
ALLIANCE

SEG

SOUTHERN
EXAMINING
GROUP

GCSE Science

Single Award and Double Award

Name ....

School .... Class / Set ....

Stanley Thornes (Publishers) Ltd

First published in 1998 by:
Stanley Thornes (Publishers) Ltd
Ellenborough House
Wellington Street
Cheltenham GL50 1YW
England

98 99 00 01 / 10 9 8 7 6 5 4 3 2 1

A catalogue record of this book is available from the British Library.

ISBN 0-7487-3668-9

Typeset by Mathematical Composition Setters, Salisbury, Wiltshire
Printed and bound in Spain by Mateu Cromo
Artwork by Peters & Zabransky

Contents

Introduction

This book has been written by examiners and endorsed by SEG for their Modular Science syllabuses (Double Award 2630 or Single Award 2620) and their Non-Modular Science syllabuses (Double Award 2610 and Single Award 2600). Please ask your teacher which syllabus you are following. The book will guide you through your syllabus and help you to prepare more effectively for tests and the examination at the end of your course.

This book follows the content of all SEG syllabuses precisely. It covers all the topics of Double Science. If you are studying Single Science you will find the parts of this book that you have to study are clearly marked at the start of each topic. Remember that for Higher Tier you can be tested on the whole contents of this book. If you are preparing for Foundation Tier, you do not have to do sections marked with **H**.

Revision notes

This book is divided into 33 topics: 11 in Sc2 (Biology), 11 in Sc3 (Chemistry) and 11 in Sc4 (Physics). These topics have been grouped in the order of the Modular syllabuses (see Contents on the previous page).

In each topic you will find a relevant photograph and a list of 'Key Words'. These are the science words you need to know and understand for this topic. If you do not know the meaning of any of these words you can look up the word in the Glossary (pages 188−94).

Work your way through the Revision Notes section. Here there are gaps for you to fill in using the Key Words (You may need to use a plural version of a word in the list). Alternatively, you may be given a choice of words, e.g. *increases/ decreases/stays the same*. You have to select the best answer from the list. When you have completed the Revision Notes you can check your answers with your teacher or use the answers in this book.

When you have worked through all of the topics you will have a concise record of the important knowledge for your SEG syllabus topics.

For high grades in GCSE Science you should be able to write chemical equations using symbols. To help you to do this there is a section on Formulae and Equations on pages 160−62.

If you are doing a SEG Modular syllabus then 25% of the marks are awarded for 9 Module Tests during the course. Another 25% comes from Coursework during the course. The final 50% of the total marks comes from an examination at the end of the course.

To help you prepare for the Module Tests there are Multiple Choice Questions on pages 163−87.

Summary questions for GCSE

The examination tests other skills as well including data handling, calculations, drawing conclusions and evaluation. These skills are best developed by trying examination questions especially in the six months before you take the examination. This book contains questions like the ones you will face in the examination.

Some will ask you to recall knowledge and show understanding. Others will test the other skills. They include questions where you need to make longer answers and show a good element of planning in your answer. Your teacher may have a copy of the answers and tips to show how you can improve your answers.

We hope that this book will help you with your course and revision in the run-up to the examination. SEG are pleased to support you through their endorsement of and involvement with this publication. We wish you success.

Bob McDuell
Keith Hirst
Graham Booth

Acknowledgements

The authors and publisher would like to thank the following for supplying photographs:

Britstock-IFA: p. 44 (TPL)
Brookes & Vernons PR: p. 140
Dr Olaf Linden/ICCE: p. 49
Francis Gohier/Ardea: p. 24
GeoScience Features: pp. 82, 112, 150
Heather Angel: p. 108
Holt Studios International: p. 122 (Andy Burridge)
Martyn Chillmaid: pp. 41, 54, 77, 117
Rotary Burnand: p. 126
Science Photo Library: p. 14 (Dick Luria); p. 92 (Biophoto Associates)

Life processes and blood

Do all of this topic for Single Science and Double Science.
H = for Higher Tier only

📝 Revision notes

Life processes

> 🔑 excretion   growth   movement   nutrition   reproduction
> respiration   sensitivity

The chick has the following life processes in common with all living organisms:

● Obtaining food by eating plants or by eating other animals is [1] _____ .

● Releasing energy from food is [2] _____ .

● Releasing waste products is [3] _____ .

● Producing offspring is [4] _____ .

● Developing from a chick to an adult is [5] _____ .

● Reacting to the surroundings is [6] _____ .

● Changing position is [7] _____ .

Functions of cell parts

> 🔑 cell membrane   chromosome   cytoplasm   nucleus
> *mitochondria*

Complete this table showing the functions of the parts of cells:

function	part of cell
Contains [8] _____ which carry genes controlling the cell's characteristics	[9] _____
Most of the chemical reactions occur here	[10] _____
Controls the movement of substances in and out of the cell	[11] _____ _____
Energy is released here	[H12] _____

Groups of cells

> **organ**   **organ system**   **tissue**

A group of cells with a similar structure and a particular function is called a [13] _____ .

Groups of tissues are called an [14] _____ .

Different organs working together form an [15] _____ _____ .

Circulation

> **aorta**   **capillary**   **pulmonary artery**   **pulmonary vein**   **vena cava**

Label the types of blood vessel on this diagram of the circulatory system:

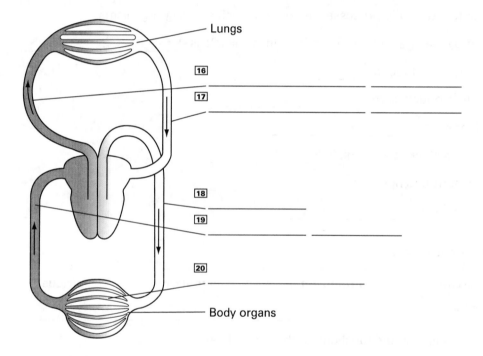

Lungs

[16] _____ _____

[17] _____ _____

[18] _____

[19] _____ _____

[20] _____

Body organs

Transport by the blood

> **carbon dioxide**   **haemoglobin**   **hormone**   **oxygen**   **soluble food**
> **urea**

Blood consists of red cells, white cells and platelets carried in a liquid called plasma.

Red cells are biconcave in shape and contain a red pigment called [21] _____ .

Their function is to transport [22] _____ .

Complete this table for transport of substances by plasma:

substance in plasma	transported from	transported to
23 _____ _____	organs	lungs
24 _____ _____	small intestine	liver
25 _____	liver	kidney
26 _____	ductless glands	organs

Defence by the blood

clot	ingest
antibody	*antitoxin*

White blood cells protect the body in two ways. Some types can 27 _____ microbes.

Others produce H28 _____ which kill microbes. Others produce

H29 _____ to counteract poisons produced by microbes.

Platelets protect the body by helping to form a 30 _____ at the site of a wound to prevent

the entry of microbes.

Summary questions for GCSE   H = for Higher Tier only

31 Complete the table for the life processes of the eagle.

Activity	Life process
eating rabbits	
flying	
laying eggs	
breathing out	
releasing energy from food	
blinking in bright sunlight	

[Total 6 marks]

32 The drawing shows a sperm cell.

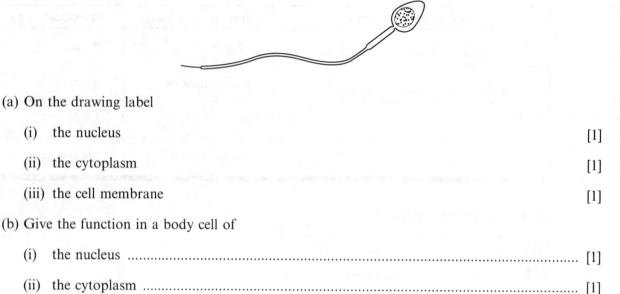

(a) On the drawing label

    (i) the nucleus   [1]

    (ii) the cytoplasm   [1]

    (iii) the cell membrane   [1]

(b) Give the function in a body cell of

    (i) the nucleus .... [1]

    (ii) the cytoplasm .... [1]

    (iii) the cell membrane .... [1]

    [H](iv) the mitochondria .... [1]

(c) The job of the sperm is to swim to find an egg. Give **two** ways its structure helps it to do this job.

    1 ....

    2 .... [2]

[Total 9 marks]

33 The drawing shows human blood seen through a microscope.

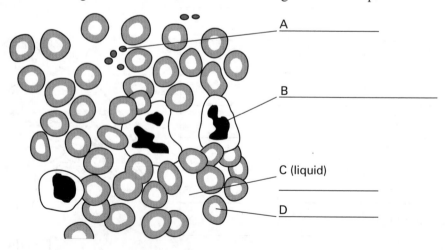

(a) On the drawing, name the parts of blood labelled **A**, **B**, **C** and **D**.   [4]

(b) Give **one** function of

    (i)   red cells .... [1]

    (ii) white cells .... [1]

    (iii) plasma .... [1]

    (iv) platelets .... [1]

[Total 8 marks]

H34 Red blood cells are involved in the transport of gases used and produced during respiration. Explain how the structure of a red blood cell makes it efficient for carrying these gases.

....

....

....

....

....

....

[Total 5 marks]

H35 Describe how the white blood cells help to defend the body against microbes.

....

....

....

....

....

....

[Total 5 marks]

Digestion

Do all of this topic for Single Science and Double Science.
H = for Higher Tier only

 Revision notes

Teeth

🔑	canine	incisor	molar	premolar

Label the types of teeth shown in the drawing of the mouth.

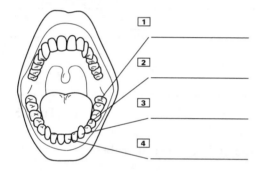

1 _____

2 _____

3 _____

4 _____

Incisor teeth have a ⑤ *sharp/pointed/flat* shape for cutting food. Canine teeth have a ⑥ *sharp/pointed/flat* shape for tearing food. Premolar and molar teeth have a ⑦ *sharp/pointed/flat* shape for grinding food.

Structure of the digestive system

🔑	gall bladder	large intestine	liver	oesophagus	pancreas
	small intestine	stomach			

Label the parts of the digestive system.

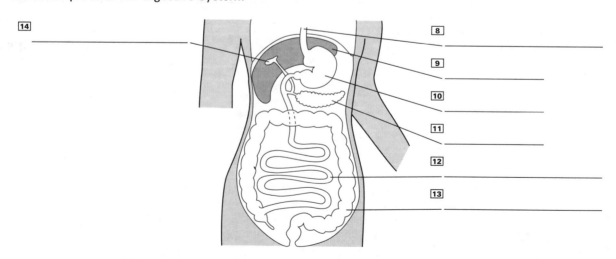

14 _____

8 _____

9 _____

10 _____

11 _____

12 _____

13 _____

Digestion and absorption

amino acids	amylase	faeces	fatty acids	hydrochloric acid
insoluble	insulin	salivary gland	soluble	sugar
bile	*glycogen*	*lipase*	*protease*	*surface area*
villi				

We need the digestive system to break down large ⑮ _____ food molecules into smaller ⑯ _____ molecules that can be absorbed into the bloodstream.

Saliva is produced by the ⑰ _____ _____ . Saliva contains an enzyme called ⑱ _____ . This enzyme breaks down starch into ⑲ _____ .

The stomach produces an enzyme called H20 _____ which breaks down protein into ㉑ _____ _____ . The stomach also produces ㉒ _____ _____ . The pancreas produces an enzyme called H23 _____ which breaks down fats into ㉔ _____ _____ . The pancreas also makes a hormone called ㉕ _____ . To help the digestion of fats, the liver produces H26 _____ which is stored in the gall bladder and which breaks down fats into droplets.

The inner surface of the small intestine is folded and has finger-like projections called H27 _____ which increase the H28 _____ _____ for the absorption of soluble food. They are also thin and well supplied with blood vessels. The liver stores excess sugar as H29 _____ . It also removes poisons such as alcohol from the blood. Indigestible food passes from the small intestine into the large intestine where most of the water is absorbed from it, forming ㉚ _____ .

Summary questions for GCSE   H = for Higher Tier only

31   The drawing shows the digestive system.

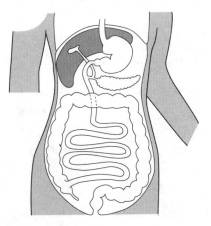

(a) On the drawing, label

    ᴴ(i)   an organ that produces protease enzymes   [1]

    ᴴ(ii) an organ that produces lipase enzymes   [1]

    (iii) where most of the absorption of soluble foods occurs   [1]

    (iv) where faeces are formed.   ... [1]

(b) Explain how the shape of the incisor teeth and molar teeth makes them suitable for different functions.

    (i) Incisors ....

    ....

    ....

    (ii) Molars ....

    ....

    .... [4]

[Total 8 marks]

H32   A student set up the experiment shown in the diagram.

- Elastic band
- Dialysis tube
- Protein and protease mixture
- Boiling tube
- Water

Explain how and why the composition of the water in the test tube would have changed one hour after the experiment had been set up.

....

....

....

....

[Total 4 marks]

H33 The drawing shows the inside of the small intestine.

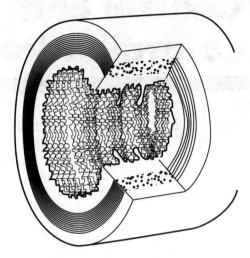

Use information from the drawing to help you explain how the structure of the lining of the small intestine makes it efficient for absorbing soluble food.

....

....

....

....

[Total 4 marks]

H34 Describe the role of the liver and the pancreas in the digestion of fats.

....

....

....

[Total 4 marks]

Control and co-ordination

 Revision notes

Receptors

| ciliary muscle | cornea | ear | eye | iris | lens |
| nose | optic nerve | pupil | retina | skin | |

Cells called receptors detect stimuli (changes in the environment). Different parts of the body contain different receptors. Complete this table about receptors:

receptor cells	found in
light	1 _____
sound and balance	2 _____
temperature and touch	3 _____
chemicals	tongue and 4 _____

The eye contains receptor cells. Label this drawing of a section through the eye:

5 _____

6 _____

7 _____

8 _____

9 _____

10 _____

11 _____

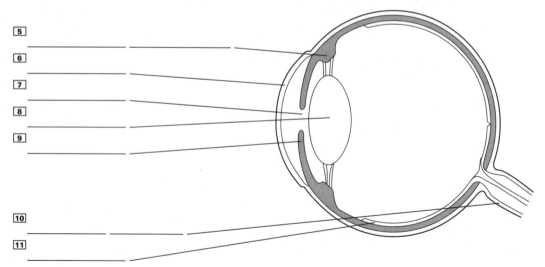

Transmission of information and control

| connector | effector | impulse | neurone | reflex action | sensory |
| synapse |

When stimulated, receptors produce nerve [12] _____ . These pass along nerve cells

called [13] _____ .

Nerve cells that carry impulses towards the brain or spinal cord are called [14] _____

neurones. Nerve cells that carry impulses within the brain and spinal cord are called

[15] _____ neurones. Nerve cells that carry impulses away from the brain or spinal

cord are called [16] _____ neurones. Impulses pass by chemical means between neurones

across junctions called [H17] _____ .

Automatic control of an activity is called a [18] _____ _____ .

The muscle or gland which brings about this response is called the [19] _____ .

The eye and light

| ciliary muscle | cornea | iris |

The amount of light entering the eye is controlled by responses of the [20] _____ .

Light is focused on the retina by the lens and the [21] _____ . The shape of the lens can be

altered by the contraction or relaxation of the [22] _____ _____ .

Co-ordination

| diabetes | endocrine | glucose | hormone | insulin |
| oestrogen | testosterone |

Chemical messengers are called [23] _____ . They are made in

[24] _____ glands and transported to their target organs by blood plasma.

The testes produce [25] _____ which controls physical changes in boys during

adolescence.

The ovaries produce [26] _____ , which controls physical changes in girls during

adolescence. Manufactured hormones are used to control female fertility.

The pancreas secretes a hormone called [27] _____ which controls the level of

[28] _____ in the blood. If insufficient insulin is made a disease called

[29] _____ results, one effect of which is [30] *high/low* levels of glucose in the blood.

🔲 **Summary questions for GCSE**   **H = for Higher Tier only**

31   The drawing shows the head of a horse. Different parts of the head contain receptors sensitive to different stimuli.

Label with an **A** part of the head which contains mainly receptors sensitive to sound.   [1]

Label with a **B** part of the head which contains mainly receptors sensitive to chemicals.   [1]

Label with a **C** part of the head which contains mainly receptors sensitive to light.   [1]

[Total 3 marks]

32   The drawing shows the front of the eye.

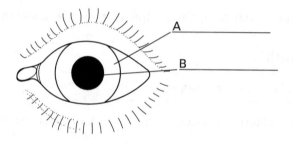

(a) Label parts **A** and **B**.   [2]

(b) How would the appearance of the front of the eye change if a bright light were shone at the eye?

.... [2]

(c) The diagram shows a section through the eye and the rays of light **X** and **Y**. The path of ray **Y** has been completed for you. Complete the path of ray **X** through the eye.   [2]

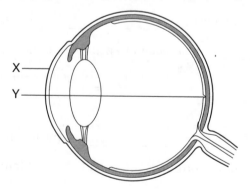

[Total 6 marks]

33 The drawing shows the structures in a pain-withdrawal reflex. When the person steps on the pin the muscle automatically contracts.

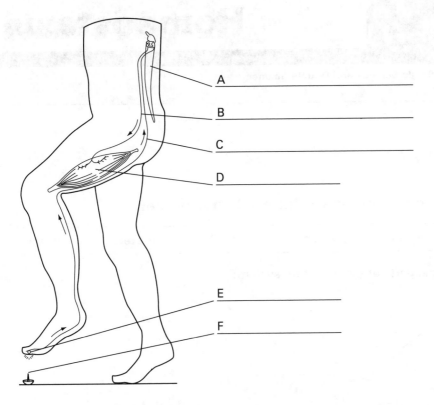

A _____

B _____

C _____

D _____

E _____

F _____

Use words from the list to label the parts **A**, **B**, **C**, **D**, **E** and **F**.

muscle   effector neurone   sensory neurone   spinal cord   stimulus   receptor

[6]

[Total 6 marks]

34 (a) Where is insulin produced? .... [1]

(b) What stimulates the production of insulin? .... [1]

(c) What effect does insulin have on the body? .... [1]

(d) Name the disease produced by insufficient secretion of insulin. .... [1]

[Total 4 marks]

Homeostasis

Do all of this topic for Single Science and Double Science.
H = for Higher Tier only

 Revision notes

Organs that maintain a constant internal environment

artery	bladder	kidney	renal	ureter	urethra

Label this drawing of the parts of the excretory system:

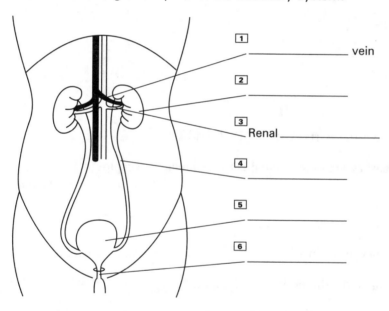

1. _____ vein

2. _____

3. Renal _____

4. _____

5. _____

6. _____

Waste materials

bladder	skin	sweat	ureter	urethra	urine

The waste liquid produced by the skin is called [7] _____ .

The waste liquid produced by the kidneys is called [8] _____ .

Urine is produced in the kidneys, then passes down the [9] _____ , and then to the

[10] _____ where it is stored. It leaves the body via the [11] _____ .

Excess heat is transferred from the body mainly via the [12] _____ .

How the kidneys work

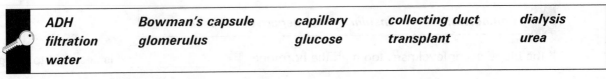

ADH	Bowman's capsule	capillary	collecting duct	dialysis
filtration	glomerulus	glucose	transplant	urea
water				

Label the drawing of the nephron.

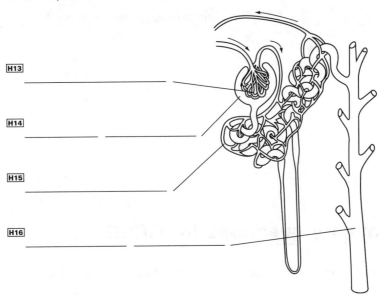

H13 _____

H14 _____ _____

H15 _____

H16 _____ _____

The first stage in urine production is the H17 _____ of blood under high pressure.

As the filtrate flows through the kidney tubules all of the H18 _____ is re-absorbed into the

blood. The ions and H19 _____ needed by the body are also re-absorbed. The remaining

filtrate, called urine, consists mainly of excess water, excess ions and H20 _____ .

If the water content of the blood is too low the hormone H21 _____ is secreted.

This H22 *increases/decreases* the rate of water re-absorption in the kidneys. Kidney failure may be

treated by using a H23 _____ machine or by means of a kidney

H24 _____ .

Temperature regulation

constrict	dilate	evaporate

Body temperature is monitored and controlled by the brain. If the body is too warm, blood vessels

supplying the skin capillaries 25 _____ to increase the blood flow through the capillaries.

If the body is too cold these blood vessels 26 _____ to reduce the flow of blood

through the skin capillaries. Increased sweating cools the body as the sweat

27 _____ .

Blood sugar regulation

homeostasis	insulin	negative feedback

If the blood sugar level rises too high the hormone [H28] _____ is secreted to bring it back

to the normal level. Maintenance of a steady internal environment is known as

[H29] _____ . The control mechanisms rely on [H30] _____

_____ mechanisms.

Summary questions for GCSE H = for Higher Tier only

31 The drawing shows some of the organs concerned with excretion.

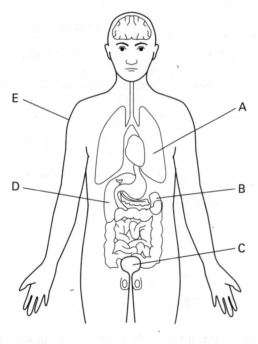

(a) Which organ (**A**–**E**)

(i) produces sweat? .... [1]

(ii) gets rid of most carbon dioxide? .... [1]

(iii) produces urine? .... [1]

(iv) stores urine? .... [1]

(b) On hot days we usually drink more than we do on cold days. Explain why.

....

....

.... [3]

[Total 7 marks]

H32 Long-distance runners are faced with particular problems on hot humid days.

(a) The runners produce far more sweat on a humid day than when it is not humid. Suggest an explanation for this.

....

....

....

....

.... [4]

(b) The loss of sweat can rise to as much as 3 litres per hour.

Explain how mechanisms inside the body try to maintain a constant amount of water in the blood in spite of the loss of this amount of sweat.

....

....

....

....

.... [6]

(c) Explain why it is important to keep a constant amount of water in the blood.

....

....

....

.... [2]

[Total 12 marks]

33 The table shows the amounts of glucose and insulin in the blood of a healthy person over a five hour period. The person ate a meal rich in starch one hour after measurements began, then rested for the remainder of the time.

Time in hours	Glucose concentration in plasma in mg per 100 cm^3	Insulin concentration in plasma in arbitrary units
0	60	10
1	60	10
2	110	60
3	130	80
4	100	10
5	60	10

(a) (i)   Where is insulin produced? .... [1]

(ii)   What stimulates the production of insulin? .... [1]

H(b) Describe the mechanisms which resulted in the rise and fall of glucose levels in the blood plasma.

....

....

....

....

....

....

.... [5]

[Total 7 marks]

Reproduction, inheritance and health

Do all of this topic for Single Science and Double Science.
H = for Higher Tier only

Revision notes

Genetic information

> **chromosome**　　　**nucleus**
>
> *allele*

Genetic information is located inside the ☐1 _____ of a cell. Genes are carried on

☐2 _____ . Many genes have two forms called ᴴ³ _____ . In body cells

the chromosomes are ᴴ⁴ *single/in pairs* whereas in gametes they are ᴴ⁵ *single/in pairs*.

Gender determination

In human body cells one pair of chromosomes carries the genes which determine sex. These are

called sex chromosomes and there are two types: X chromosomes and Y chromosomes.

Complete this diagram to show how gender is determined in humans. Use 'X' for an X chromosome

and 'Y' for a Y chromosome.

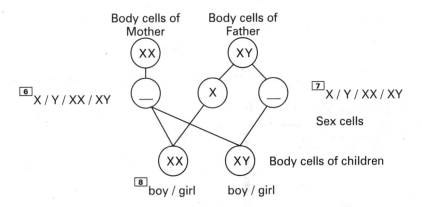

Body cells of Mother (XX)　Body cells of Father (XY)

☐6 X / Y / XX / XY

☐7 X / Y / XX / XY

Sex cells

(XX)　(XY)　Body cells of children

☐8 boy / girl　boy / girl

Genetic terms

> *dominant*　　*genotype*　　*heterozygous*　　*homozygous*　　*recessive*

An allele which controls the development of a characteristic when it is present on only one of the chromosomes is a [H9] _____ allele. An allele which controls the development of a characteristic only when it is present on both of the chromosomes is a [H10] _____ allele. If both chromosomes in a pair contain the same allele of a gene the individual is [H11] _____ for that gene. If the chromosomes in a pair contain different alleles of a gene, the individual is [H12] _____ for that gene. The composition of the alleles of an individual is known as the [H13] _____ .

Inheritance of a disorder caused by a dominant allele

Mendel studied inheritance in peas. One characteristic he studied was height. If pure-breeding tall pea plants were crossed with each other, all the F_1 plants were tall. If these F_1 plants were crossed with each other, some of the F_2 plants were tall and some short. Complete the Punnett square to show the chance of an F_2 plant being tall.

	alleles in pollen	
alleles in ovule	T	t
T	[H14] *TT/Tt/tt*	Tt
t	[H15] *TT/Tt/tt*	tt

The probability of the offspring being short is [H16] *nil/evens/1 in 2/1 in 4.*

Inheritance of a disorder caused by a recessive allele

Cystic fibrosis is caused by a recessive allele c. It can be passed on by parents even though neither of them is affected.

Complete the Punnett square diagram to show the probability of a child inheriting the disorder from a heterozygous father (Cc) and a homozygous mother (CC).

	alleles in sperm	
alleles in egg	C	c
C	CC	Cc
C	CC	[H17] *CC/Cc/cc*

The probability of the child inheriting the disorder is [H18] *nil/25%/50%/75%/100%.*

Cell division and genetic engineering

assortment	bacteria	DNA	meiosis	mitosis

The cells of offspring produced by asexual reproduction are produced by [H19] _____ from

the parental cells. Sex cells are produced by [H20] _____ from the parental cells.

During meiosis the number of chromosomes in a cell [H21] *halves/doubles/stays the same*.

Sexual reproduction gives rise to variation because:

● gametes are produced by [H22] _____ ;

● during production of gametes there is random [H23] _____ of chromosomes.

The molecule that carries genetic information is called [H24] _____ . It carries this

information as a code for the sequence of amino acids in proteins. In genetic engineering a sequence

of this molecule is cut out from a host cell and transferred into [H25] _____ which then

divide rapidly and make large quantities of the substance coded for.

Drugs

addiction	brain	cancer	cilia	depressant

Solvents, tobacco smoke and alcohol can all affect our behaviour. The craving for drugs or solvents is

known as [26] _____ . Sniffing solvents is most likely to cause damage to the lungs,

the kidneys and the [27] _____ . Tobacco smoke contains substances which can cause

lung [28] _____ . It also prevents the action of [29] _____ which line the breathing

passages. These contain a mucous membrane which makes the air moist and keeps the lungs clean.

Smoking also produces carbon monoxide which reduces the oxygen-carrying capacity of the fetus.

One result of this is to reduce the birth mass of the fetus. Alcohol affects the nervous system by

slowing down our reactions and it is therefore known as a [30] _____ .

Summary questions for GCSE   H = for Higher Tier only

31 Complete the sentences.

(a) Characteristics such as eye colour are controlled by parts of chromosomes called

.... [1]

(b) The sex chromosomes in the body cells of a boy are .... [1]

(c) The sex chromosomes in the body cells of a girl are .... [1]

[Total 3 marks]

32 The drawing is from a health education pamphlet.

ONE UNIT	ONE UNIT	ONE UNIT	ONE UNIT	ONE UNIT
$\frac{1}{2}$ PINT OF ORDINARY STRENGTH BEER LAGER OR CIDER 3.5% ABV†	1 SMALL GLASS OF WINE 11% ABV†	1 SINGLE MEASURE* OF SPIRITS 40% ABV†	1 SMALL GLASS OF SHERRY 16% ABV†	1 SINGLE MEASURE* OF APERITIF 15% ABV†

*This applies to the $\frac{1}{6}$ gill measure in most of England and Wales. In N. Ireland a pub measure is $\frac{1}{4}$ gill and in Scotland $\frac{1}{5}$ or $\frac{1}{6}$ gill. † These strengths (Alcohol by Volume) appear on all labels

(a) Which of the drinks contains the highest concentration of alcohol? .... [1]

(b) A partygoer drinks 4 pints of beer and 2 double measures of spirits. How many units of alcohol does he consume?

Answer .... units [2]

(c) Explain why it would be dangerous for the partygoer to drive home.

....

.... [2]

[Total 5 marks]

H33 (a) A human skin cell has 46 chromosomes. When it divides:

(i) how many chromosomes will each daughter cell have? .... [1]

(ii) which type of cell division is involved? .... [1]

(b) (i) How many chromosomes does a sperm-producing cell in the human testis have? .... [1]

(ii) How many chromosomes does a human sperm have? .... [1]

(iii) Which type of cell division is involved in the production of sperm? .... [1]

(c) Explain why sexual reproduction leads to variation in the offspring but asexual reproduction does not.

....

....

....

.... [4]

[Total 9 marks]

H34 Use a genetic diagram to explain how the disorder cystic fibrosis can be inherited from parents neither of whom has the disorder.

[Total 5 marks]

H35 Sickle-cell anaemia is a disorder of red blood cells caused by a recessive allele **h**. The diagram shows the inheritance of sickle cell anaemia in a family.

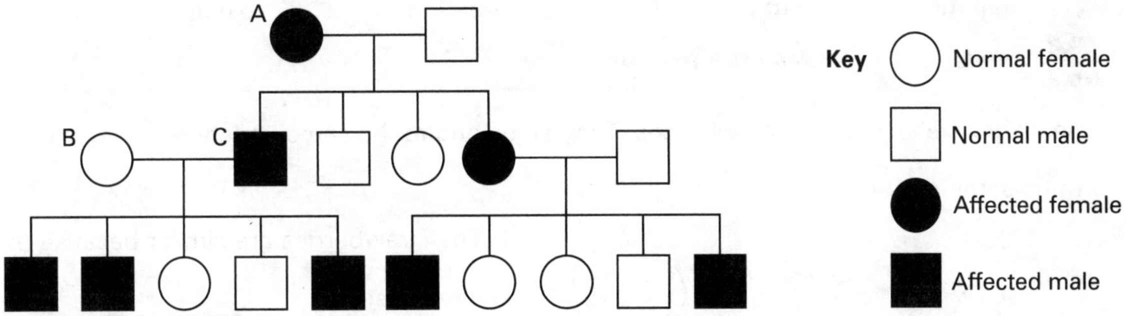

(a) Which alleles are present in the body cells of

(i) **A**? .... [1]

(ii) **B**? .... [1]

(iii) **C**? .... [1]

(b) **B** and **C** decide to have another child.

What is the chance that this child will inherit the condition? .... [1]

Use a genetic diagram to explain your answer. [4]

[Total 8 marks]

Growth, variation and evolution

Do all of this topic for Single Science and Double Science.
H = for Higher Tier only

 Revision notes

Variation

asexual	cancer	characteristic	chromosome	clone
continuous	discontinuous	environmental	fertilisation	gene
genetic	mutation	radiation	sexual	
allele	*selective breeding*			

The diagram shows some strawberries. They all belong to the same variety and they were all picked on the same day.

The strawberries are similar because they all contain similar [1] _____ . Any variations in size are due to

[2] *e*_____ factors and

[3] *g*_____ factors.

This type of variation is known as

[4] _____ variation because there is a wide range of different sizes. Where individuals can be placed in only two groups the variation is said to be [5] _____ .

New strawberry plants are produced when the parent plant forms runners. Runners are produced by cell division at the base of the parent cell. This type of reproduction is known as [6] _____ reproduction. Because of this the genetic information in the cells of the parent plant and the young plant is [7] *identical / different*.

Strawberry fruits are produced when sex cells from two strawberry flowers fuse. The process by which two sex cells fuse is called [8] _____ and this method of reproduction is said to be [9] _____ . The fruits produced by this method of reproduction contain

[10] *identical/different* genetic information. Genetically identical organisms are called

[11] _____ .

Genes are carried on [12] _____ . Many genes have different forms called

[H13] _____ . New forms of genes result from changes, called [14] _____ ,

to existing genes. The frequency of these changes is increased by exposure to ionising

[15] _____ and to certain chemicals. Some mutations in body cells cause

uncontrolled cell division which may result in [16] _____ .

We can use [H17] _____ _____ to produce new varieties of plants and

animals. We do this by choosing for breeding individuals that have [18] _____ that

are useful to us.

Growth

breed	**fertile**	**specialisation**	
mitosis			

Growth involves cell division, cell enlargement and cell [19] _____ .

The cells of offspring produced by asexual reproduction are produced by [H20] _____ from

the parental cells. Each of these cells has [H21] *the same number/half the number/double the number*

of chromosomes that the parent cell has.

Members of the same species are able to [22] _____ and produce [23] _____

offspring.

Evolution

decay	**evolution**	**fossil**	**variation**
advantageous	*Darwin*	*Lamarck*	

The remains of organisms which lived many years ago are called [24] _____ . They are most

often formed from parts of organisms that do not [25] _____ . Most of the evidence for

[26] _____ comes from a study of the way in which organisms have changed over

long periods of time.

The theory of evolution by natural selection states that all members of a species show

[27] _____ . Some variations are [H28] _____ and enable the

organism to survive changing conditions, whereas other organisms become extinct.

The theory of evolution was proposed by H29 _____ . This superseded a theory of

evolution by the inheritance of acquired characteristics proposed by H30 _____ .

Summary questions for GCSE   H = for Higher Tier only

31 A class collected 200 limpet shells from a rocky shore and measured their lengths. The results are shown in the table.

Length of shell in mm	8–11	12–15	16–19	20–23	24–27	28–31
Number of limpets	14	30	50	46	44	16

(a) Plot a bar chart of the results.

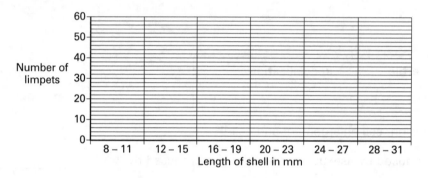

[2]

(b) Suggest **two** reasons for the variation in shell length of the limpets.

1 .... [1]

2 .... [1]

[Total 4 marks]

32 The drawings show five stages in the possible evolution of the horse. The drawings are to the same scale. All the animals except *Equus* are now extinct and all are herbivores.

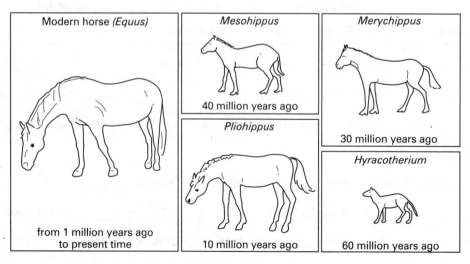

(a) Explain how we know that *Hyracotherium* lived 60 million years ago.

....

....

.... [3]

(b) Describe **two** ways in which horses seem to have changed over the last 60 million years.

1 .... [1]

2 .... [1]

(c) Suggest **one** reason why *Merychippus* became extinct.

....

.... [1]

ᴴ(d) Use information from the drawings to suggest an explanation for the way in which *Equus* could have evolved.

....

....

....

....

....

.... [6]

[Total 12 marks]

H33   The table shows changes in the average annual milk yield for a dairy cow in England.

Year	Average milk yield per cow in litres
1938	2547
1948	2834
1958	3390
1968	3709

(a) By how much did the average milk yield per cow rise between 1938 and 1968? .... [1]

(b) Explain how farmers achieved this increase in milk yield from these dairy herds.

....

....

.... [3]

[Total 4 marks]

The environment

Do all of this topic for Single Science and Double Science.
H = for Higher Tier only

 Revision notes

Adaptions for survival

camouflage	insulation	surface area	wax

The polar bear lives in the arctic. Its white colour

gives it [1] _____ against the

background of snow. It has a thick layer of fat

under its skin for [2] _____ .

It has a [3] *thick/thin* coat of fur and [4] *large/small*

ears to reduce heat loss.

The cactus lives in a hot desert. Its leaves are shaped like spines to reduce

the [5] _____ _____ through which water can

evaporate. Its surface is covered by a thick layer of [6] _____ to

reduce evaporation.

Competition, predators and prey

breeding	nutrient	predator	prey

Plants compete with each other for light, water and nutrients. Animals compete with each other for

[7] _____ and space for [8] _____ .

Animals that hunt and eat other animals are called [9] _____; the animals that are

eaten are called [10] _____ . If the number of prey rises, the number of predators will

usually [11] *rise/fall*; if the number of predators rises the number of prey will usually [12] *rise/fall*.

Human populations and air pollution

acidic	carbon dioxide	combustion	fossil fuel	leaves
non-renewable	pollution	sulphur dioxide		

Increases in the size of human populations have led to increased use of [13] _____

energy resources such as [14] _____ _____ . The burning of fuels is known as

[15] _____ ; this leads to [16] _____ of the air.

Power stations affect the environment. The most abundant gas in the smoke from power stations is

[17] _____ _____ . There are also significant amounts of the gas

[18] _____ _____ which dissolves in rain making it [19] _____ .

When this rain falls it causes the trees to lose some of their [20] _____ .

When this rainwater reaches the lake it causes the lake to become [21] _____ , killing many

of the organisms that live there.

Effects of agriculture

erosion	food chain
eutrophication	*fertiliser*

Farmers use [H22] _____ to replace nutrients in the soil. If fertilisers or sewage

reach fresh water, water plants grow rapidly and as a result [H23] _____ occurs

resulting in oxygen depletion and the death of many of the animals. Farmers use pesticides to kill

organisms which damage crops. The use of these disrupts [24] _____ _____ .

The use of larger fields for growing crops causes soil [25] _____ .

The greenhouse effect and the ozone layer

carbon dioxide	CFC	methane	radiation	ultraviolet

Increases in the number of cattle and in the number of rice fields have resulted in an increase in the

amount of the gas [26] _____ in the atmosphere. Deforestation has reduced the rate at which

[27] _____ _____ is removed from the atmosphere and locked up in wood.

These two gases reduce the amount of energy lost from the Earth by [28] _____ ,

causing the mean temperature of the Earth to rise. This effect is known as global warming.

The ozone layer absorbs harmful [29] _____ radiation. Increased use of chemicals such

as [30] _____ is producing 'holes' in the ozone layer leading to increased risk of skin cancer.

Summary questions for GCSE   H = for Higher Tier only

31 An ornamental pond contains aquatic plants, small animals and fish. The small animals feed on the plants and the fish feed on the small animals. The pond is 'balanced', i.e. it was set up several years ago and now needs very little attention. The numbers of each type of animal and plant stay relatively constant.

(a) Suggest **two** factors that limit the number of plants growing in the pool.

1 .... [1]

2 .... [1]

(b) Suggest **two** factors that limit the number of small animals growing in the pool.

1 .... [1]

2 .... [1]

(c) Explain in terms of predators and prey why the numbers of small animals and fish stay fairly constant.

....

....

.... [3]

[Total 7 marks]

32 (a) Explain how acid rain is formed.

....

....

.... [4]

(b) The chart shows the damage to trees caused by acid rain.

(i)   What percentage of pine trees had no damage? .... [1]

(ii) What percentage of spruce trees were slightly damaged? .... [1]

(c) Suggest **one** explanation for the fact that many fir trees were badly damaged, but none of the oak trees in the same forest were badly damaged.

....

.... [2]

[Total 8 marks]

H33 A paper manufacturer is building a factory on the banks of a river. The effluent from the factory will pass into the river. This effluent will contain carbohydrates.

Explain the effects the effluent will have on life in the river in the first few weeks.

....

....

....

....

....

[Total 5 marks]

H34 In some lands to the south of the Sahara desert, forests are being destroyed to provide land for crops. The technique is called 'slash and burn' – the trees are hacked down and burned. Crops can be grown successfully on this land for a few years, but then crop yields fall and the land is abandoned as semi-desert.

(a) Explain the effects on the atmosphere of removing large areas of forest by the 'slash and burn' method.

....

....

....

.... [4]

(b) Explain why the cleared land soon becomes semi-desert, and suggest what could be done to prevent this.

....

....

....

.... [4]

[Total 8 marks]

Particles and the Periodic Table

Do all of this topic for Single Science and Double Science.
H = for Higher Tier only

📄 Revision notes

Matter made up of particles

boiling point	chromatography	condensing	distillation	evaporating
filtering	fractional distillation	freezing	melting	

The diagram below shows the changes of state between solids, liquids and gases. Label the diagram using key words.

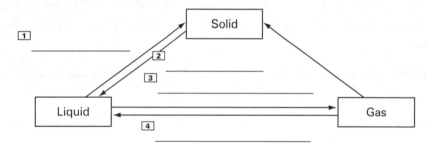

Solid

[1] _____

[2] _____

[3] _____

Liquid

[4] _____

Gas

Separating mixtures

In a mixture of sand and water, the sand can be separated by [5] _____ .

A solute can be recovered from a solution by [6] _____ .

A solvent can be recovered from a solution, e.g. water from salt solution, by

[7] _____ .

A mixture of two liquids that completely mix together can be separated by

[8] _____ _____ which uses differences in the

[9] _____ _____ of the two liquids to separate them.

A solution of dyes in water can be separated by [10] _____ .

Chemical combination

atom	chlorine	combined	compound	electron	helium
inert	nucleus	proton	shell	sulphide	

Elements are made up of tiny particles called [11] _____ . Atoms of different elements can

be joined together or [12] _____ to form a [13] _____ .

A mixture of iron and sulphur, when heated, forms iron [14] _____ .

An atom consists of protons and neutrons tightly packed in the [15] _____ with

[16] _____ moving rapidly around the nucleus in certain shells.

The nucleus of an atom is [17] *positively charged/negatively charged/neutral*.

A neutral atom contains equal numbers of [18] _____ in the nucleus and

[19] _____ outside the nucleus.

The element which can be used to fill the airship in the photograph is [20] _____ . Unlike

hydrogen, this gas is very unreactive or [21] _____ .

An element used to make household bleaches to kill germs is [22] _____ .

The Periodic Table

alkali	alkali metal	atomic mass	atomic number	group
halogen	hydrogen	hydrogen chloride	metal	noble gas
non-metal	period	shell	trend	
chlorine	*iodine*	*sodium hydroxide*		

A horizontal row in the Periodic Table is called a [23] _____ and a vertical column is called

a [24] _____ . The elements in Mendeléev's original Periodic Table were arranged in order

of increasing [25] _____ _____ but in the modern Periodic Table they are

arranged in order of increasing [26] _____ _____ .

In any period of the Periodic Table, the elements change across the period from [27] _____

on the left-hand side to [28] _____ on the right-hand side. Elements in the same

group are similar but not identical. There are patterns or [29] _____ within each group.

Elements in group 1 of the Periodic table are called the [30] _____ _____ and

elements in group 7 are called the [31] _____ . The unreactive elements in group 0 are

called [32] _____ _____ .

Lithium is an alkali metal.

Lithium reacts with water to form lithium hydroxide and [33] _____ .

When lithium hydroxide is tested with universal indicator, it turns purple, showing that lithium

hydroxide is an [34] _____ .

Bromine is an element in group 7 of the Periodic Table. Bromine will displace [H35] _____

from potassium iodide solution but not [H36] _____ from potassium chloride solution.

Hydrochloric acid is made by dissolving [37] _____ _____ gas in water.

Electrolysis of brine (sodium chloride solution) produces hydrogen gas, [H38] _____ gas

and [H39] _____ _____ solution.

Elements in the same group each contain the same number of electrons in the outer

[40] _____ .

Summary questions for GCSE   H = for Higher Tier only

[41]   The table gives information about four elements in group 1 of the Periodic Table.

Element	Symbol	Atomic number	Melting point in °C	Boiling point in °C	Density in g/cm³	Reaction with cold water
lithium	Li	3	180	1347	0.53	steady reaction
sodium	Na	11	98	883	0.97	fast reaction
potassium	K	19	64	774	0.86	very fast reaction hydrogen catches alight
rubidium	Rb	37	39	688	1.53	

(a) What pattern links the atomic number with the melting and boiling points of elements in group 1?

.... [1]

(b) Which alkali metal in the table sinks when added to water?

.... [1]

(c) From your knowledge of the properties of the alkali metals, suggest what you would see when a lump of rubidium is freshly cut.

....

.... [2]

(d) (i) Predict the reaction of rubidium with cold water.

.... [1]

(ii) Write a balanced symbol equation for the reaction of rubidium with cold water.

.... [2]

[Total 7 marks]

H42 Chlorine gas is bubbled into solutions of potassium fluoride, potassium chloride and potassium iodide.

The table shows some observations made.

Solution	Appearance before bubbling chlorine through	Appearance after bubbling chlorine through
potassium fluoride	colourless solution	colourless solution
potassium chloride	colourless solution	colourless solution
potassium bromide	colourless solution	red solution
potassium iodide	colourless solution	brown solution

(a) Why did no reaction take place when chlorine was passed through potassium fluoride solution?

....

.... [1]

(b) What name is given to the type of reaction taking place when chlorine is bubbled through potassium bromide or potassium iodide solutions?

.... [1]

(c) Write word, symbol and ionic equations for the reaction taking place when chlorine is bubbled through potassium bromide solution.

Word ....

Symbol ....

Ionic .... [5]

[Total 7 marks]

H43 Explain, in terms of atomic structure, why

(a) potassium is more reactive than sodium.

....

.... [2]

(b) bromine is less reactive than chlorine.

....

.... [2]

(c) argon is unreactive.

....

.... [2]

(d) the atomic radii of group 2 elements increase down the group.

....

.... [2]

(e) the energy required to remove one electron from a potassium atom is relatively small but the energy required to remove two is very much higher.

....

....

.... [3]

[Total 11 marks]

Reactivity series

Do all of this topic for Single Science and Double Science.
H = for Higher Tier only

Revision notes

Metals and the reactivity series

oxygen	paraffin oil	reactivity

Metals are arranged in order of [1] _____ with the most reactive metals at

the [2] *bottom/top* of the list. Metals at the top of the list, such as potassium and sodium, are stored

under [3] _____ _____ to prevent them reacting with [4] _____,

water and carbon dioxide in the air.

Reactions of metals

copper	displacement	hydrogen	hydroxide	iron	oxide
magnesium	salt	zinc			

When a metal reacts with oxygen in the air it will form an [5] _____ . When a metal reacts

with water it will form a metal [6] _____ and [7] _____ gas. If a metal

reacts with a dilute acid, a metal [8] _____ is formed and [9] _____ gas. One

metal which does not react with water or dilute hydrochloric acid is [10] _____ .

Reactions of metals can be predicted using the reactivity series. If a metal is added to a solution of a

less reactive metal which is in a compound, a reaction will take place. This type of reaction is called a

[11] _____ reaction.

If iron filings are added to blue copper(II) sulphate solution, a brown deposit of

[12] _____ and a pale green solution of [13] _____ (II) sulphate are formed.

Two metals which will displace lead from a solution of lead(II) nitrate are [14] *z* _____ and

[15] *m* _____ .

Oxidation and corrosion

copper	corrosion	electrolysis	iron(III) oxide	magnesium
oxidation	reduction	rust	uncombined	zinc
electron	*oxide*			

The photograph shows a rusting ship. The rusting of iron and steel is a [16] _____

process. Iron reacts with oxygen and water to form a red-brown solid called [17] _____

which is hydrated [18] _____ _____ . A process where oxygen is added or

where [19] _____ are lost is called [20] _____ . A process where oxygen

is lost is called [21] _____ .

Rusting of iron and steel can be slowed down by sacrificial protection when a reactive metal such as

magnesium or [22] _____ is in contact with the iron or steel.

A metal low in the reactivity series, such as [23] _____ , is unreactive and corrodes

[24] *slowly/quickly*. A metal high in the reactivity series, such as [25] _____ , will

corrode [26] *slowly/quickly*. A metal high in the reactivity series is extracted from its ores

by [27] _____ . A metal such as iron, in the middle of the reactivity series, is

extracted from its ores by [28] _____ . A metal such as gold, at the bottom of the

reactivity series, may be found [29] _____ in the Earth.

Aluminium is a reactive metal but it does not rapidly corrode because of the formation of an insoluble,

impermeable layer of aluminium [H30] _____ .

Summary questions for GCSE

The reactivity series is shown below. Use this to answer any of the questions in this section.

potassium K	**most reactive**
sodium Na	
calcium Ca	
magnesium Mg	
aluminium Al	
zinc Zn	
iron Fe	
hydrogen H	
copper Cu	
silver Ag	
gold Au	**least reactive**

31 (a) Explain why a reaction between zinc and silver nitrate solution will take place.

....

.... [2]

(b) Complete and balance the symbol equation for this reaction.

$Zn(s) + ... AgNO_3(aq) \rightarrow$ .... + ... [3]

(c) You have all of the metals in the reactivity series and solutions of their nitrates.

Explain how you would try to find the position of the metal nickel in the reactivity series.

....

....

.... [3]

[Total 8 marks]

32 The table gives the results of the reactions of four metals, **W**, **X**, **Y** and **Z**, with air, water and dilute hydrochloric acid.

Metal	Reaction with air	Reaction with cold water	Reaction with acid
W	reacts with air on heating	no reaction	few bubbles of gas
X	no reaction on heating	no reaction	no reaction
Y	reacts with air without heating	reacts steadily	rapid production of gas bubbles
Z	reacts with air on heating	very slow reaction	steady production of gas bubbles

(a) (i) Name the gas produced when a metal reacts with dilute hydrochloric acid.

.... [1]

(ii) How would you carry out a positive test for this gas? Give the result of this test

....

.... [2]

(b) Put these four metals, **W**, **X**, **Y** and **Z**, in the correct order of reactivity (most reactive first).

.... [3]

[Total 6 marks]

33 Four experiments, labelled **A–D**, were carried out to find the conditions needed for iron to rust.

Four test tubes, each containing an iron nail, were sealed so no more air could enter them.

The tests tubes are shown in the diagram.

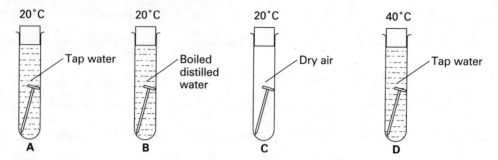

The test tubes were left for two weeks.

The results are shown in the table.

Test tube	What was seen after two weeks
A	rust
B	no rust
C	no rust
D	rust

(a) What do experiments **A**, **B** and **C** show about the rusting of iron?

.... [2]

(b) The nail in **D** rusted faster than the nail in **A**.

After two weeks the same amount of rust was seen in test tubes A and D.

Explain these observations.

....

.... [2]

(c) Four ways of protecting iron from rusting are

galvanising   **greasing**   **painting**   **sacrificial protection**

Finish the table to show the best method of preventing the steel objects from rusting.

One has been done for you. Each method should be used once.

Steel object	Method of protection
car body panel	painting
blade of a saw	
leg of a pier always under water	
wire fence	

[2]

[Total 6 marks]

Acids, bases and salts

Do all of this topic for Single Science and Double Science.
H = for Higher Tier only

📝 Revision notes

Acids, alkalis and pH

🔑	acid	alkali	hydrogen	neutral	salt
	sour	strong	universal	weak	

The photograph shows substances which contain [1] _____ . They have a

[2] _____ taste. All acids contain the element [3] _____ . When a metal reacts

with an acid a [4] _____ and hydrogen are formed. Substances with pH values less than 7

are [5] _____ . Substances with pH values greater than 7 are [6] _____ .

A substance with a pH value of exactly 7 is [7] _____ .

The pH of a solution can be found by using [8] _____ indicator. A solution with a

pH value of 8 is a [9] _____ alkali and a solution with a pH value of 1 is a

[10] _____ acid.

Reactions of acids

🔑	carbon dioxide	hydrogen	limewater	salt	sulphate

When magnesium ribbon is added to an acid, bubbles of colourless [11] _____ gas are

seen. When sodium carbonate crystals are added to an acid, bubbles of colourless

[12] _____ _____ gas are seen. This gas turns [13] _____

milky.

When dilute sulphuric acid is warmed with black copper(II) oxide, a blue solution of copper(II)

[14] _____ is formed.

Mixing an alkali with an acid produces water and a [15] _____ .

Equations

🔑 | alkali | ammonia | carbon dioxide | nitric | salt | sodium
| sulphate | sulphuric | water

Complete the following word equations.

Acid + [16] _____ → [17] _____ + water

Sodium hydroxide + [18] _____ acid → [19] _____ nitrate + water

Zinc oxide + [20] _____ acid → zinc [21] _____ + [22] _____

[23] _____ + sulphuric acid → ammonium sulphate

Calcium carbonate + hydrochloric acid → calcium chloride + [24] _____ _____

+ water

Zinc + [25] _____ acid → zinc [26] _____ + hydrogen

📖 Summary questions for GCSE **H = for Higher Tier only**

27 Here are the names of some salts. Use this list to answer the questions which follow.

barium chloride   **barium carbonate**   **barium sulphate**   **calcium nitrate**
calcium sulphate   **lead(II) nitrate**   **lead(II) sulphate**   **ammonium sulphate**

(a) Which salt does **not** contain a metal?

.... [1]

(b) (i) Which salt reacts with dilute hydrochloric acid to produce bubbles of colourless gas?

.... [1]

(ii) Which gas is produced?

.... [1]

(c) Which salt contains only two elements?

.... [1]

[Total 4 marks]

28 Copper(II) sulphate crystals can be made by reacting copper(II) oxide with sulphuric acid. The equation for the reaction is:

$CuO(s) + H_2SO_4(aq) \rightarrow CuSO_4(aq) + H_2O(l)$

Write an account of how a sample of copper(II) sulphate crystals could be made from copper(II) oxide.

....

....

....

....

....

....

....

[Total 5 marks]

H29 Sodium burns in oxygen to form sodium oxide. Sodium oxide is a base.

(a) What is a base?

.... [1]

(b) If a base dissolves in water it forms an alkali.

(i) Complete an ionic equation for the reaction of solid sodium oxide with water.

.... + $H_2O(l) \rightarrow$ .... [2]

(ii) Write an ionic equation for the reaction of sodium hydroxide with dilute hydrochloric acid.

.... [2]

(c) Explain why the energy released when 1 mole of sodium hydroxide and 1 mole of hydrochloric acid are mixed is the same as when 1 mole of potassium hydroxide and 1 mole of nitric acid are mixed.

....

.... [1]

[Total 6 marks]

Rates of chemical reactions

Do all of this topic for Single Science and Double Science.

 Revision notes

Rates of reactions

catalyst	collision	explosion	reactant
chlorine	concentrated	pressure	surface area

The photograph shows a chemical reaction which is finished in a fraction of a second. It is a [1] *very fast/very slow* reaction. It is called an [2] _____ . The erosion of stonework on a building is a [3] *very fast/very slow* reaction. As the time for a reaction increases the rate of the reaction [4] *decreases/increases/stays the same*.

Rates of reaction can be explained using [5] _____ theory.

Chemical reactions can be speeded up in a number of ways.

Increasing the temperature of a reaction will [6] *slow down/speed up* a chemical reaction. Heating up the reaction mixture [7] *speeds up/slows down* the particles. This will produce more effective [8] _____ .

Increasing the concentration of one or more of the reactants will [9] *increase/decrease* the time taken for the reaction and so will [10] *increase/decrease* the rate of reaction. In the case of a mixture of gases reacting, increasing the [11] _____ of the gases is the same as increasing the concentrations. Increasing the concentration results in more [12] _____ between reacting particles and more of these will be effective, leading to a faster reaction.

A solid in the form of a powder reacts [13] *faster than/slower than/at the same rate as* the same mass of the solid in the form of lumps. This is because the powder has a larger [14] _____ _____ to come in contact with the other reactant. Although coal does not readily catch light, mixtures of coal dust and air can cause an [15] _____ in a coal mine.

A mixture of hydrogen and chlorine reacts only very slowly but explodes in sunlight. Sunlight provides energy to start the reaction by breaking some of the bonds between pairs of [16] _____ atoms.

The rate of a chemical reaction can be changed by using a [17] _____ whose mass [18] *stays the same/increases/decreases* throughout the reaction. Using a catalyst produces [19] *more product/less product/the same mass of product* at the end of the reaction. Catalysts only operate in particular reactions.

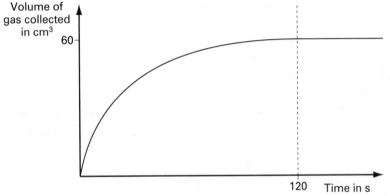

The graph shows the results obtained in a reaction in which calcium carbonate lumps and dilute hydrochloric acid react.

The graph is level when the reaction is [20] *fast/slow/finished* and one of the [21] _____ has been used up. In the early stages of the reaction the graph is steepest as the reaction is [22] *fastest/slowest* at this stage because the acid is most [23] _____. If the experiment was repeated with the same mass of powdered calcium carbonate, the volume of carbon dioxide collected would be [24] *the same/more than 60 cm³/less than 60 cm³*. The reaction is finished in [25] *less than/more than/exactly* two minutes.

✎ Summary questions for GCSE

[26] Ben carried out a series of experiments to compare the rates of reaction between sodium thiosulphate and dilute hydrochloric acid under different conditions.

$$Na_2S_2O_3(aq) + 2HCl(aq) \rightarrow 2NaCl(aq) + S(s) + H_2O(l) + SO_2(g)$$

He carried out the reactions in 100 cm³ glass beakers – either a squat beaker or a tall beaker. The beaker was stood on a piece of paper with a cross on it. The time was taken until the cross disappeared from view when viewed from above.

(a) Why does the cross on the paper disappear from view?

.... [1]

Here is the table he produced on which to record his results.

Experiment number	Volume of water in cm³	Volume of hydrochloric acid in cm³	Volume of sodium thiosulphate in cm³	Temperature in °C	Type of beaker	Time in s
1	30	5	30	20	squat	
2	20	5	30	50	squat	
3	20	5	30	20	squat	
4	40	5	10	20	squat	
5	10	5	40	20	squat	
6	20	10	30	20	squat	
7	30	5	20	20	squat	
8	20	5	30	40	squat	
9	20	5	30	30	squat	
10	20	5	30	20	tall	

(b) In which order should the water, dilute hydrochloric acid and sodium thiosulphate solutions be mixed?

.... [1]

(c) When should the temperature of each experiment be taken?

.... [1]

(d) At which point should he have started timing each experiment?

.... [1]

(e) Which experiment would you expect to be fastest? Explain your choice.

....

.... [2]

(f) (i) Which **four** experiments would best show the effect of changes in the concentration of sodium thiosulphate solution on the rate of reaction?

.... [1]

(ii) Predict the pattern of the results of these experiments you would expect. Use your knowledge of particles and their collisions to explain your prediction.

....

.... [3]

(g) (i) Which **four** experiments would best show the effect of increasing temperature on the rate of reaction?

.... [1]

(ii) Predict the pattern of the results of these experiments you would expect. Use your knowledge of particles and their collisions to explain your prediction.

....

....

.... [3]

(h) Explain why you would expect the time for Experiment 10 to be less than the time for Experiment 3.

....

.... [2]

[Total 15 marks]

27 The equation for the decomposition of hydrogen peroxide is:

$2H_2O_2(aq) \rightarrow 2H_2O(l) + O_2(g)$

The reaction is very slow but is catalysed by many substances including manganese oxide.

(a) Name the gas produced on the decomposition of hydrogen peroxide solution.

.... [1]

(b) An experiment was carried out using the apparatus in the diagram.

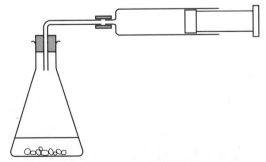

10 cm³ of hydrogen peroxide was placed in the flask and 0.5 g of manganese oxide added. The rubber bung was quickly replaced and the volume of gas collected was measured at 30 second intervals. The results are shown in the table.

Time in s	0	30	60	90	120	180	210	240
Volume of gas in cm³	0	27	45	57	64	68	70	70

(i)   Draw a graph of the volume of gas collected against time.

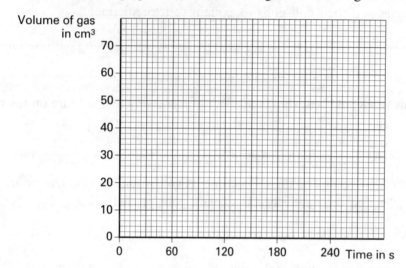

[3]

(ii) What mass of manganese oxide would remain at the end of the experiment?

.... [1]

(iii) The experiment was repeated with another metal oxide. This metal oxide was a better catalyst than manganese oxide. On the same grid sketch a graph you would expect for the decomposition of another sample of hydrogen peroxide with the other metal oxide.

[2]

[Total 7 marks]

Chemicals from oil

Do all of this topic for Single Science and Double Science.
H = for Higher Tier only

 ## Revision notes

Carbon compounds

gas	organic	petroleum

Compounds of carbon, excluding simple compounds such as carbon dioxide and compounds such as sodium carbonate, are called [1] _____ compounds. Crude oil

(or [2] _____) is a mixture of complex hydrocarbons found in the Earth. It was

formed over millions of years when high temperatures and pressures acted on the remains of tiny sea

creatures. It is often found with natural [3] _____ .

Fossil fuels

carbon	detergent	fossil	gas
impermeable rock	oxygen	water	

Label this diagram showing where oil and gas are found in the Earth.

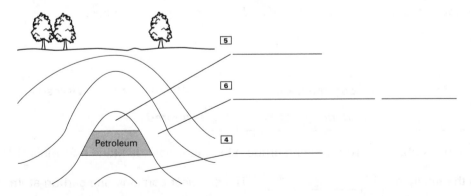

Crude oil, natural gas and coal are all [7] _____ fuels. These fuels were formed from living

materials which decayed in the absence of [8] _____ under high temperatures and

pressures. These fuels contain the element [9] _____ either as the element or combined.

The photograph shows crude oil leaking from a tanker. This can cause environmental disasters. Chemicals used to disperse oil spills are called [10] _____ .

Oil refining

🔑	boiling point	crude oil vapour	diesel oil	fractional distillation
	fuel oil	paraffin	petrol	refinery gases

Crude oil is refined by a process of [11] _____ _____ . This involves splitting up the mixture into different fractions, each fraction having a different range of [12] _____ _____ . The diagram shows a column used for refining crude oil. Label this diagram.

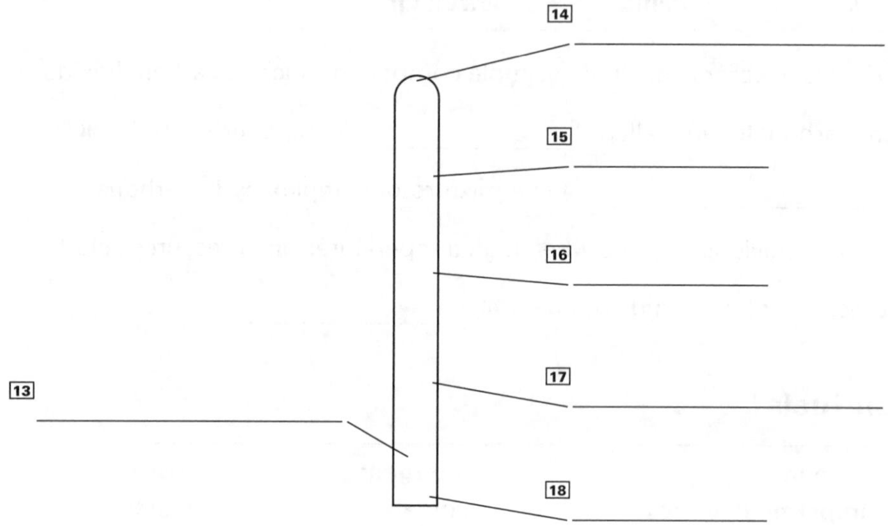

[14] _____

[15] _____

[16] _____

[13] _____

[17] _____

[18] _____

Fractions with a range of low boiling points condense near the [19] *top/middle/bottom* of the tower. As you go down the tower the fractions have [20] *higher/lower* boiling point ranges.

Alkanes

🔑	carbon dioxide	carbon monoxide	methane	viscosity
	alkane	*covalent*	*saturated*	

Most of the compounds obtained from refining crude oil are compounds fitting a formula C_nH_{2n+2}. These belong to the family of [H21] _____ . The simplest contains one carbon atom and is called [22] _____ . It is the major constituent of natural gas. All alkanes contain only single [H23] _____ bonds between carbon atoms. They are said to be [H24] _____ compounds.

Alkanes are generally unreactive compounds. As the number of carbon atoms in the alkane increases the boiling point of the alkane [25] *increases/decreases*, the resistance to pouring or

[26] _____ increases and the alkane is [27] *easier/more difficult* to burn. Hydrocarbons burn in air or oxygen. If they are burned in an excess of air, [28] _____

_____ and water are produced. If they are burned in a limited supply of air, a poisonous

gas called [29] _____ _____ may be formed in addition to water.

Cracking hydrocarbons

catalyst	cracking
alkene	*unsaturated*

High boiling point fractions are of less economic value. They are often split up by a process

called [30] _____ into smaller molecules. This process involves passing the vapour of the

high boiling point fraction over a heated [31] _____ . The smaller molecules produced often

contain hydrocarbons containing a double bond between carbon atoms. These are called

[H32] _____ compounds. They belong to the family of hydrocarbons called

[H33] _____ .

Alkenes and polymers

ethene				
1,2-dibromoethane	*ethene*	*monomer*	*polymer*	*polymerisation*

The simplest alkene has a formula C_2H_4 and is called [34] _____ . It reacts with a solution of

bromine to form a [H35] *green/colourless/clear/brown* compound called 1,2-dibromoethane. The

word equation for this reaction is:

[H36] _____ + bromine → [H37] _____

Small unsaturated molecules can be joined together by a process of addition

[H38] _____ to form long chains called [H39] _____ . The small molecules

are called [H40] _____ .

Summary questions for GCSE   H = for Higher Tier only

41 The table gives some information about five hydrocarbons found in crude oil.

Hydrocarbon	Boiling point in °C	State at room temperature	Number of carbon atoms	Formula
pentane	40	liquid	5	C_5H_{12}
hexane	70	liquid	6	C_6H_{14}
octane	125	liquid	8	
nonane		liquid	9	C_9H_{20}
decane	175	liquid	10	$C_{10}H_{22}$

(a) (i) Put a ring round the most likely boiling point of nonane.

      50 °C   100 °C   150 °C   200 °C   [1]

  (ii) The table shows the pattern between the number of carbon atoms in the hydrocarbon and the boiling point.

     What is the pattern?

     .... [1]

(b) What is the formula of octane? .... [1]

(c) Which hydrocarbon is shown in the formula below?

```
    H   H   H   H   H   H
    |   |   |   |   |   |
H – C – C – C – C – C – C – H
    |   |   |   |   |   |
    H   H   H   H   H   H
```
                                .... [1]

(d) Complete the word equations for the burning of pentane in air.

    pentane + air (plentiful supply) → .... + ... [1]

    pentane + air (limited supply) → .... + ... [1]

[Total 6 marks]

H42 Liquid paraffin is a mixture of alkanes. The alkanes in the mixture have between 10 and 20 carbon atoms.

A reaction takes place when liquid paraffin vapour is passed over heated broken china.
The apparatus in the diagram can be used for this.

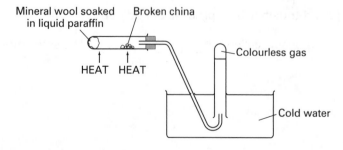

A colourless gas is collected in test tubes over water.

(a) Why are the properties of the first test tube of gas collected different from those of all of the others collected?

.... [1]

(b) What can you conclude about the solubility of the colourless gas in water?

.... [1]

The gas collected is ethene, C_2H_4. The structure of ethene is

```
   H        H
    \      /
     C == C
    /      \
   H        H
```

(c) Complete the balanced symbol equation for the combustion of ethene in a plentiful supply of air.

.... C_2H_4 + ... $O_2 \rightarrow$ .... + ... H_2O [2]

(d) Ethene can be polymerised by passing it over a heated catalyst.

(i) What is the name of the polymer formed? .... [1]

(ii) Draw the structure of part of the polymer chain formed.

[2]

(e) This polymer is used to make plastic toys for young children and for making twine for gardeners to tie up their plants. Give **two** reasons why this polymer is suitable for each use.

Children's toys:

1 .... [1]

2 .... [1]

Gardener's twine:

1 .... [1]

2 .... [1]

[Total 11 marks]

Using electricity

Do all of this topic for Single Science and Double Science.
H = for Higher Tier only

Revision notes

Circuits and resistance

ammeter	component	current	energy	light	ohm
parallel	series	voltage	voltmeter		

Electric current is measured in amps (A) using an [1] _____ .

The job of the current in a circuit is to transfer [2] _____ from the power supply to the circuit components. Energy can be transferred to heat, [3] _____ and movement in circuit components.

A circuit that has only one path for the current is a [4] _____ circuit. Where there are two or more possible current paths the circuit is a [5] _____ one.

The current is the same at all points in a series circuit; none is gained or lost. The supply [6] _____ is shared between the components. Components connected in parallel have the same [7] _____ across them. The total current is the sum of the currents in the individual [8] _____ .

Voltage is measured using a [9] _____ that is placed in [10] _____ with a power supply or component.

The amount of current passing in a circuit depends on the [11] _____ and the resistance of the circuit. Increasing the voltage causes the current to increase but increasing the resistance causes the current to decrease.

The voltage across a circuit component is equal to the current times its resistance, $V = I \times R$.

Resistance is measured in [12] _____ (Ω). Provided that the temperature stays the same, the resistance of a metal wire does not change when the current changes. This graph of current against voltage is a [13] *curve/straight line* passing through the origin.

A diode allows [14] _____ to pass in only one direction, shown by the direction of the arrow on the circuit symbol.

The resistance of a light-dependent resistor (LDR) and a thermistor depend on environmental conditions. The resistance of an LDR decreases as the light level [H15] *decreases/increases* and that of a thermistor [H16] *decreases/increases* as the temperature increases.

Electricity in the home

alternating	circuit breaker	conductor	current	earth	energy
fuse	insulated	insulation	kilowatt	kilowatt-hour	live
neutral	power	resistance	voltage	watt	

The current due to a cell or battery in a circuit is always in the same direction; it is called a direct current (d.c.). A current that changes direction is called an [17] _____ current (a.c.). Mains electricity uses alternating current that changes direction 100 times each second.

The three conductors that form the electricity cable to a house are called the [18] / _____, [19] *n*_____ and [20] *e*_____. Energy is supplied through the [21] _____ wire. The [22] _____ wire is the return path that completes the circuit and the [23] _____ wire is for safety. When an appliance is operating normally there is no current in the [24] _____ wire.

A metal-cased appliance such as a toaster needs all three conductors. The flexible cable that connects the toaster to the mains supply has three separate wires, each of which has a layer of [25] _____.

The wire that has blue insulation is the [26] _____, the wire with brown insulation is the [27] _____ and that with green and yellow insulation is the [28] _____.

At the toaster, the live and [29] _____ are connected to the heating element and the [30] _____ is connected to the metal case. The switch and a [31] _____ are also connected in the live conductor as shown in the diagram.

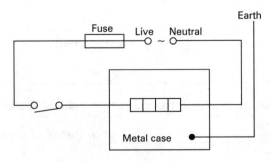

The fuse protects against a fire hazard in the connecting wires. If a fault in the element causes too large a [32] _____ the fuse melts and breaks the circuit. The fuse and the [33] _____ wire together protect the user from electrocution. If the case becomes live the earth wire provides a low [34] _____ path to earth. This causes a large current and the fuse melts, breaking the circuit.

Appliances such as televisions and hairdriers normally have [35] *metal/plastic/wooden* cases which cannot become live because this material is not [36] _____ . These appliances are said to be double [37] _____ and they do not need an [38] _____ wire.

In addition to the fuse fitted to the plug of each appliance, each circuit in a house has its own fuse or [39] _____ _____ to protect the fixed cables from overheating and causing a fire. Circuit breakers are more reliable than [40] _____ and they are easily reset when the fault has been put right.

Fuses and circuit breakers have a current rating. If the current exceeds this then the circuit is broken. The rating of a fuse or circuit breaker is normally the next higher available value above the normal current.

All electrical appliances have a power rating. This is the rate at which [41] _____ is transferred from the electricity supply. Power is measured in [42] _____ or kilowatts and can be calculated using the equation *power = current* $\times$ [43] _____ or $P = I \times V$.

The total energy transfer while an appliance is operating can be calculated using the equation *energy =* [44] _____ $\times$ *time*. Energy is calculated in joules when the power is in watts and the time is in seconds.

Electricity boards charge us for the energy that is supplied to our homes. The joule is too small a unit to use, so the electricity meters in houses measure the energy transfer in units of [45] _____ (kW h). This energy transfer can be calculated using the energy equation above, but with the power in [46] _____ and the time in hours.

Generating electricity

electromagnetic induction **magnet** **magnetic field**	
electromagnet	

The generation and transmission of electricity depend on electromagnetism. A voltage is created in any conductor that moves through a [47] _____ _____ or that is positioned within a magnetic field that is changing in size or direction. This is known as

48 _____ _____ and can be demonstrated by using a

magnet, a coil of wire and a sensitive ammeter.

The ammeter detects a current whenever the coil or the 49 _____ is moved. The size and

direction of the current depend on the speed and direction of movement.

Sensitive ammeter

A bicycle dynamo generates electricity by rotating a magnet next to a coil of wire. A power station

generator works in a similar way; in this case an H50 _____ is rotated inside the

thick wire coils.

Summary questions for GCSE   H = for Higher Tier only

51 The diagram shows a circuit that can be used to measure the resistance of a heater.

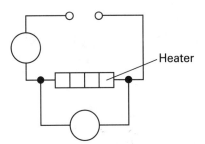

Heater

(a) Complete the circuit diagram by labelling the ammeter **A** and the voltmeter **V**.   [2]

(b) What extra component could be added to allow the current in the circuit to be varied? Draw the
circuit symbol in the correct position on the diagram.   [3]

(c) The table shows the ammeter and voltmeter readings for a range of currents.

Voltage in volts	0.70	1.35	1.80	2.40	2.85	3.20
Current in amps	0.18	0.35	0.47	0.61	0.73	0.82

Draw a graph of voltage against current.

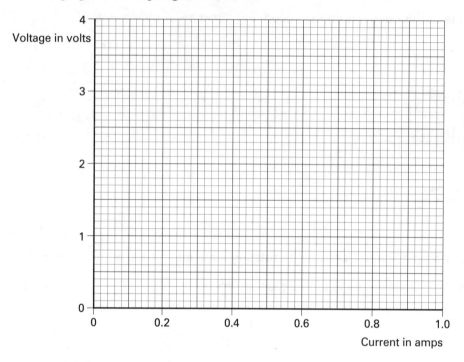

[3]

(d) Use your graph line to complete the table. [2]

Current in amps	Voltage in volts
0.20	
0.40	
0.60	
0.80	

(e) Is it true to state that 'doubling the current in the heater causes the voltage to double'? Explain your answer by referring to the data in (d).

....

.... [2]

(f) Calculate the resistance of the heater when the current passing is 0.5 A.

....

.... [3]

[Total 15 marks]

52 There are three conductors in the mains supply cable: live, neutral and earth. Which of these

(a) supplies energy to a house?

.... [1]

(b) is a safety wire?

.... [1]

(c) is the return path for the current?

.... [1]

(d) should fuses and switches be connected in?

.... [1]

[Total 4 marks]

53 Metal-cased appliances should have an earth wire connected to the casing.

(a) What hazard does the earth wire protect against?

.... [1]

(b) What else is used in a circuit to protect the user from this hazard?

.... [1]

(c) Describe what happens if the live wire touches the metal casing of an earthed appliance.

....

.... [2]

(d) Explain why an appliance that is double insulated does not need an earth connection.

....

.... [2]

[Total 6 marks]

54 Diagram **A** shows an electric radiant heater. It has a heating element and a switch. A fuse is fitted to the plug.

A

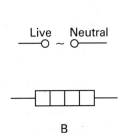

B

(a) Complete the circuit diagram (**B**) to show how the heating element is connected to the mains supply. Include the switch and fuse in your circuit. [2]

(b) (i) Write down the name of the wire that is not shown on the circuit diagram.

.... [1]

(ii) Whereabouts on the fire should this wire be connected?

.... [1]

(c) When operating from the 240 V mains supply, the current in the heating element is 4.5 A. Calculate the power of the heater.

....

....

.... [3]

[Total 7 marks]

55 Two coils of wire are wound on a soft iron core. One coil is connected to a cell and a switch. The second coil is connected to a centre-zero ammeter.

When the switch is closed, the ammeter pointer moves to the right and then back to zero.

(a) Explain why the ammeter pointer moves when the switch is closed.

....

....

.... [3]

(b) Explain why the ammeter pointer moves back to zero.

....

.... [2]

(c) Describe and explain what happens to the ammeter pointer when the switch is opened.

....

....

.... [3]

[Total 8 marks]

Force and motion

Do all of this topic for Single Science and Double Science.

Revision notes

Graphs of motion

air resistance	balanced	direction	distance
gradient	gravitational	speed	terminal velocity
weight			

The average speed of a moving object is calculated using the equation:

① _____ = $\dfrac{distance\ travelled}{time\ taken}$.

A distance–time graph shows the total ② _____ travelled by an object at each point of its

motion. The slope or ③ _____ of the graph at any point represents the

object's ④ _____ .

If you walk to the shop and then return home, the ⑤ _____ you have travelled increases as

you walk to the shop and as you return home. The graph shows how the distance travelled changes on

such a journey.

Distance

Time

The gradient of a distance–time graph can only be positive. The steeper the ⑥ _____ , the

faster the ⑦ _____ it represents.

Velocity is the speed in a given ⑧ _____ . It can have both positive and negative

values to represent motion in opposite ⑨ _____ .

All objects close to the Earth are affected by its [10] _____ pull. This force is called an object's [11] _____ .

There are two forces acting on an object falling through the air; the [12] _____ acts downwards and the [13] _____ _____ acts upwards. While the weight force remains constant, the air resistance increases as the [14] _____ increases. A sky diver accelerates if the [15] _____ is greater than the [16] _____ _____ and moves at a steady speed, called [17] _____ _____, when these forces are [18] _____ .

Turning forces

force	moment	pivot

A force has a turning effect, or [19] _____ , when it is applied at some distance from the turning point, or [20] _____ . The moment of a force is measured in Nm and is calculated from the equation *moment =* [21] _____ × *perpendicular distance to pivot*. The principle of moments applies to a system in equilibrium; it states that *the sum of the clockwise moments about a pivot is equal to the sum of the anticlockwise moments*.

Force and shape

elastic	elastic limit	force	Hooke's	origin
proportional	size	spring		

Forces cause objects to change shape. When the stretching [22] _____ is removed from a rubber band, it returns to its original shape; it is described as being [23] _____ .
Metal wires and [24] _____ stretch in a regular and predictable way up to a certain limiting [25] _____ .
Up to this limit, a graph of extension against stretching force is a straight line through the [26] _____ , showing that the extension is [27] _____ to the force. The material is said to follow [28] _____ law.
Increasing the force beyond this limit can take the spring or wire beyond its [29] _____ _____ so it no longer returns to its original [30] _____ .

Summary questions for GCSE

31 The graph shows how the distance travelled by a person out walking changes with time.

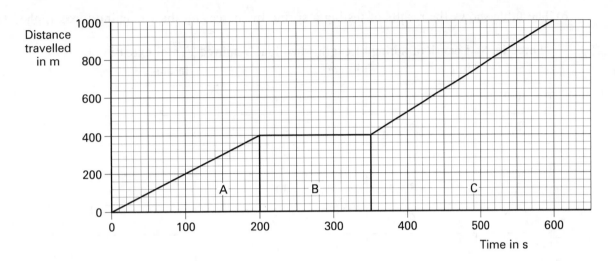

(a) How long did the walk last?

.... [1]

(b) How long did the person spend resting?

.... [1]

(c) During which lettered part of the graph did the person travel with the greatest speed? Explain how you can tell from the graph line.

....

.... [2]

(d) Calculate the speed of the person in the first 200 s of the walk.

....

....

.... [3]

(e) Calculate the average speed of the person on the whole walk.

....

.... [2]

(f) Why is it important to describe the answer to (e) as an average speed?

.... [1]

[Total 10 marks]

32   (a) Here is a list of some everyday objects.

wooden bookshelf   pencil eraser   lump of putty   sponge

Put a ring around those that are elastic.   [2]

(b) The table shows the results of testing a spring by increasing the stretching force applied to it.

Force in N	Length of spring in cm	Length when load removed in cm	Total extension in cm
0.0	4.6	4.6	0.0
0.7	5.2	4.6	
1.8	6.0	4.6	
2.7	6.8	4.6	
3.7	7.6	4.6	
4.6	8.4	4.6	
5.5	9.5	4.8	
6.1	10.2	5.1	

(i)   Complete the table by working out the total extension caused by each force applied to the spring.   [3]

(ii) Draw a graph of extension against force.   [3]

(iii) Describe what the graph shows about the effect of increasing the force on the extension of the spring.

....

....

.... [3]

(iv) For what range of forces is the extension proportional to the force? Explain how you can tell from your graph.

....

.... [2]

(v) For what range of forces is the spring elastic? Explain how you can tell.

....

.... [2]

[Total 15 marks]

Electro-magnetic radiation

Do all of this topic for Single Science and Double Science.
H = for Higher Tier only

Revision notes

Wave measurements

> amplitude   cycle   frequency   hertz   wavelength

Some measurements apply to all waves. The distance from the beginning to the end of one wave

[1] _____ is called the [2] _____ (λ) and the number of wave cycles

passing any point in one second is the [3] _____ (f). The maximum displacement

from the normal position is called the [4] _____ of a wave motion. Amplitude and

wavelength are both measured in metres (m); frequency is measured in [5] _____ (Hz).

Electromagnetic spectrum

> electromagnetic spectrum   frequency   gamma ray   infrared   light   microwave
> radio wave   ultraviolet   vacuum   wavelength   X-ray

Light is only a small part of the [6] _____ _____ , a family of waves that

all travel at the same speed in a [7] _____ . The diagram shows the complete spectrum.

Frequency in Hz

10^{20} 10^{17} 10^{14} 10^{11} 10^{8} 10^{5}

Gamma rays   Ultraviolet   Microwaves

X-rays   Infrared

Radio waves

Light

10^{-12} 10^{-9} 10^{-6} 10^{-3} 1 10^{3}
Wavelength in m

The electromagnetic spectrum is a family of waves of the same type that differ in wavelength and

[8] _____ . It extends from [9] _____ _____ , which have the

longest [10] _____ and lowest [11] _____, to X-rays and

[12] _____ _____, which have a very high [13] _____ and

short [14] _____ .

Radio and television programmes are broadcast using [15] _____ _____

while [16] _____ are used both for cooking and for radio transmissions. Television

remote controls and oven grills use [17] _____ . Our eyes detect

[18] _____ . Exposure to [19] _____ radiation can result in skin cancer,

so sunbed users need to take care.

The shortest waves, X-rays and [20] _____ _____, are very hazardous to

humans. Food and medical instruments can be sterilised using [21] _____

_____ which are also used as tracers in medicine and for treating cancer.

[22] _____ are useful for examining broken bones and for detecting flaws in other objects.

Light

| 🔑 | diffraction | reflection | refraction | sound | wavelength |

Light has a much shorter [23] _____ and travels faster than [24] _____ .

It is reflected in all directions by rough surfaces and in a predictable way by shiny surfaces and

mirrors, the angle of incidence and the angle of [25] _____ being equal.

Light changes speed and [26] _____ when it passes from one substance to

another. This is called [27] _____ and it can cause a change in direction.

The spreading out of waves as they pass through a narrow gap is called

[H28] _____ .

The diagrams on the next page show this effect at gaps of different sizes.

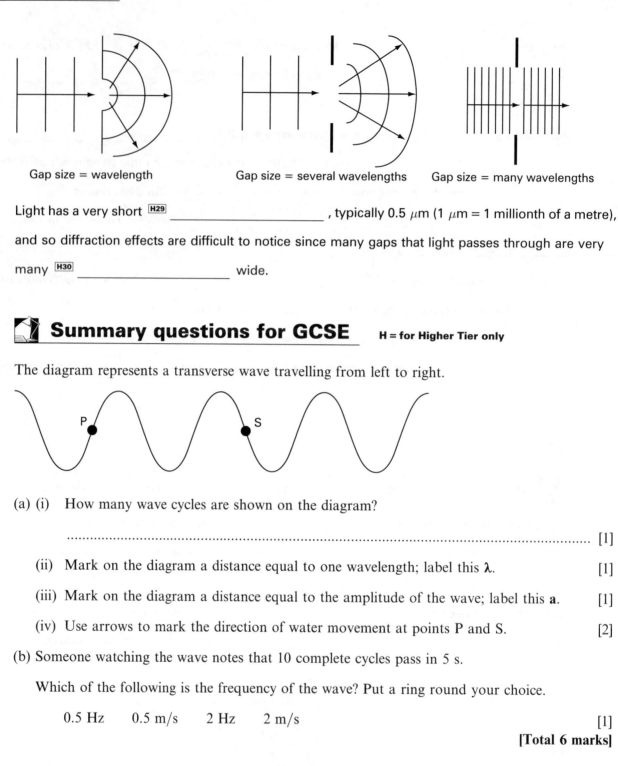

Gap size = wavelength   Gap size = several wavelengths   Gap size = many wavelengths

Light has a very short ⬛H29 _____ , typically 0.5 μm (1 μm = 1 millionth of a metre), and so diffraction effects are difficult to notice since many gaps that light passes through are very many ⬛H30 _____ wide.

Summary questions for GCSE   H = for Higher Tier only

31 The diagram represents a transverse wave travelling from left to right.

(a) (i)   How many wave cycles are shown on the diagram?

.... [1]

(ii) Mark on the diagram a distance equal to one wavelength; label this λ.   [1]

(iii) Mark on the diagram a distance equal to the amplitude of the wave; label this **a**.   [1]

(iv) Use arrows to mark the direction of water movement at points P and S.   [2]

(b) Someone watching the wave notes that 10 complete cycles pass in 5 s.

Which of the following is the frequency of the wave? Put a ring round your choice.

0.5 Hz   0.5 m/s   2 Hz   2 m/s   [1]

[Total 6 marks]

32 Here is a list of some types of wave.
gamma rays   infrared   light   microwaves   radio waves   ultraviolet

(a) Write down **two** properties that all the waves in the list have in common.

....

.... [2]

(b) Write out the list in order of increasing wavelength, starting with the shortest waves.

....

.... [3]

(c) Which **two** waves shown in the list can be used for cooking?

.... [2]

(d) X-rays are used to examine bones where there is a suspected fracture.

Photographic film is used to detect the X-rays.

(i) Describe the properties of X-rays that make it suitable for this purpose.

....

.... [2]

(ii) Explain why the people who operate X-ray equipment should avoid over-exposure and describe how this is done.

....

.... [3]

[Total 12 marks]

H33 When water waves enter a harbour through a gap in the harbour wall they can spread out.

(a) Write down the name of this effect.

.... [1]

(b) What **two** factors does the amount of spreading depend on?

.... [2]

(c) Complete the diagrams to show the amount of spreading that occurs when water waves of the same wavelength enter harbours through different-sized gaps.   [2]

[Total 5 marks]

34 The diagram represents light waves travelling from air into glass.

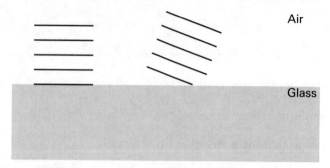

Air

Glass

(a) Complete the diagram to show the waves as they pass through the glass. [4]

(b) Choose from *decreases/stays the same/increases* to describe what happens to each of the following when the waves pass into the glass:

   (i) Wavelength .... [1]

   (ii) Frequency .... [1]

   H(iii) Speed .... [1]

[Total 7 marks]

Energy resources and energy in reactions

Do all of this topic for Single Science and Double Science.
H = for Higher Tier only

Revision notes

Energy sources

atmosphere	Earth	energy	fossil fuel	generator
hydroelectric	noise	non-renewable	oil	renewable
Sun	temperature	turbine	wave	

Most of our electricity is generated by burning [1] _____ _____ such as coal,

gas and [2] _____ . No more of these fuels can be made; they are

[3] _____ . Like most of our energy resources, the [4] _____ stored in

fossil fuels came from the [5] _____ . The energy was stored by plants through

photosynthesis.

Coal-fired and nuclear power stations use the fuel to generate steam at very high

[6] _____ and pressure. The steam loses its energy as it passes through the

[7] _____ that drive the [8] _____ , producing electricity.

The Sun, wood, wind and moving water are all examples of [9] _____ energy

resources. Energy in the wind, [10] _____ and rivers comes from the Sun heating the

atmosphere; it can be used to drive [11] _____ directly without using steam.

Wind turbines and [12] _____ power stations have the benefit of not polluting

the [13] _____ , but they are expensive to build. A wind farm takes up a large

amount of space and also causes [14] _____ pollution.

The tides provide another [15] _____ energy resource, but it is proving difficult to

obtain energy from the tides in a cost-effective way.

A tidal barrage can be used to generate electricity when the tide goes in and out; the energy in tidal

motions comes from the Moon's movement around the [16] _____ .

Storing energy

battery	efficiency	electricity	gravitational potential	kinetic
mains electricity	power	Sun		

Solar cells, which produce [17] _____ from the Sun's radiation, are also expensive

and take up a large area. They can be used on the walls of large buildings to provide some electricity

and, together with storage [18] _____ , they are useful for providing small amounts

of electricity in remote places where there is no [19] _____ _____ .

They are also useful for low [20] _____ devices such as calculators.

Solar heating uses energy from the [21] _____ directly to heat water. This can be done with

high [22] _____ by passing water through blackened pipes that absorb radiant

energy from the Sun.

Energy can be usefully stored as [23] _____ _____ energy

(gpe), e.g. in a pumped storage scheme. There is low demand for electricity at night, so it is used to

pump water from a low reservoir to a high one. At times of peak demand the water is released. As it

falls downhill, it loses [24] _____ _____ energy and gains

[25] _____ energy which is then transferred to electricity as the water passes through the

turbines.

Water released to
generate electricity
at peak demand

Water pumped to high
reservoir at night

Turbines and
pumps/generators

Burning

carbon	energy	exothermic	oxygen

The photograph shows a large forest fire. Burning is an [26] _____ reaction which

releases a large amount of [27] _____ to the surroundings. Burning involves the reaction

with [28] _____ . Wood burns to produce carbon dioxide and water. Wood **must** contain the

two elements hydrogen and [29] _____ .

Bond making and breaking

> 🔑 bond   endothermic   exothermic
>
> *bond breaking*   *bond making*

Energy is stored in 30 _____ within chemicals. When a chemical change takes place

energy changes may be observed. Reactions which give out energy to the surroundings are

called 31 _____ reactions and reactions which take in energy from the

surroundings are called 32 _____ reactions.

In a chemical reaction there is a change in bonding. Energy is required for bond breaking and is

released on H33 _____ _____ .

In an exothermic reaction less energy is required for H34 _____ _____ than

for H35 _____ _____ . In an endothermic reaction less energy is required

for H36 _____ _____ than for H37 _____ _____ .

Energy level diagrams

> 🔑 *activation energy*   *catalyst*   *energy*   *energy change*
>
> *energy level diagram*   *product*   *reactant*

The minimum energy needed by reactant particles before a reaction can occur is called the

H38 _____ _____ .

A substance which alters the rate of a reaction by lowering the activation energy is called a

H39 _____ .

The two H40 _____ _____ _____ below show exothermic and

endothermic reactions. Label the diagrams using key words.

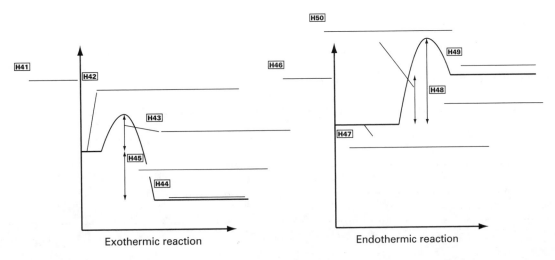

Exothermic reaction   Endothermic reaction

Summary questions for GCSE   H = for Higher Tier only

51 Here is a list of energy sources.

coal   gas   oil   waves   wind   wood

(a) Which of these are fuels?

....

.... [2]

(b) Write the name of each source in the correct column in the table.   [3]

Renewable resources	Non-renewable resources

[Total 5 marks]

52 In a pumped storage system, electricity is used to pump water from a low reservoir to a high one.

(a) Describe the energy transfer that takes place when water is pumped into the high reservoir.

....

.... [2]

(b) Explain how water in the high reservoir can be used to generate electricity.

....

....

.... [3]

(c) The efficiency of this process is 80%.

Explain what this means and suggest **two** ways in which energy is wasted.

....

....

....

.... [4]

[Total 9 marks]

53 A number of gas-fired power stations have been built in recent years to replace coal-fired power stations. They are more efficient than coal-fired power stations, they take up less space and they do not need to keep stocks of fuel.

Replacing coal-fired power stations with gas-fired ones has enabled the UK to reduce its carbon dioxide emissions.

(a) Explain why a gas-fired power station produces less carbon dioxide than a coal-fired one generating the same power.

....

....

.... [3]

(b) The known reserves of gas in the UK are much less than those of coal.

Suggest why coal-burning stations are being replaced by gas-burning ones.

....

....

.... [3]

(c) In the UK very little electricity is generated from renewable sources.

Suggest why this is the case and explain why renewable sources of energy need to be developed early in the next century.

....

....

....

.... [4]

[Total 10 marks]

The table gives the bond energies for some covalent bonds.
Use this data in the questions which follow.

Bond	Bond energy in kJ/mol	Bond	Bond energy in kJ/mol	Bond	Bond energy in kJ/mol	Bond	Bond energy in kJ/mol
H—H	436	C—H	412	Cl—Cl	242	Br—H	366
C=C	348	O—H	463	Br—Br	193	I—H	299
C—C	612	F—H	562	I—I	151	I—Cl	202
F—F	158	Cl—H	431				

H54 (a) Which bond in the table requires the most energy to break? .... [1]

(b) Which bond releases the largest amount of energy when formed? .... [1]

(c) The elements fluorine, chlorine, bromine and iodine are in Group 7 of the Periodic Table.

What is the pattern of the bond energies in F_2, Cl_2, Br_2 and I_2?

.... [1]

(d) Estimate the bond energy in the C≡C bond. .... kJ/mol [1]

(e) (i) Plot the bond energies of F—H, Cl—H and I—H against the mass of 1 mol of F—H, Cl—H and I—H.

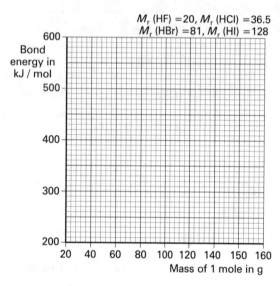

M_r (HF) =20, M_r (HCl) =36.5
M_r (HBr) =81, M_r (HI) =128

Draw the best curve. [3]

(ii) **From the graph**, estimate the bond energy of the Br—H bond.

....................kJ/mol [1]

(f) The reaction of hydrogen and bromine can be represented by the equation

$H_2(g) + Br_2(g) \rightarrow 2HBr(g)$

(i) Work out the total energy required to break 1 mol of H—H and 1 mol of Br—Br bonds.

....................kJ [1]

(ii) The energy released when 1 mol of hydrogen reacts with 1 mol of bromine is 72 kJ. Work out the bond energy of the Br—H bond.

Answer ....kJ/mol [2]

[Total 11 marks]

Transferring energy

Do all of this topic for Single Science and Double Science.

Revision notes

Energy transfer

efficiency	electricity	heat	light	movement	sound

Doing any work always involves an energy transfer. A hairdrier is designed to transfer energy from electricity into [1] h_____ and [2] m_____ of the air. It also transfers some energy into [3] _____ . A television is designed to produce [4] l_____ and sound but it also produces heat.

Kettles and immersion heaters are designed to transfer energy from electricity to [5] _____ in the water. They are very efficient at doing this. Filament lamps only transfer a small amount of the energy from [6] _____ into light; they have a low [7] _____ . Energy-efficient lamps produce the same [8] _____ output as filament lamps but take in less energy from [9] _____ .

The efficiency of a device is calculated using the relationship $efficiency = \dfrac{useful\ energy\ output}{total\ energy\ input}$.

Methods of energy transfer

absorber	density	electromagnetic radiation	emitter	energy
evaporation	expand	gas	infrared	insulator
metal	radiation	surface	temperature	

There is a flow of energy between places at different [10] _____ . This flow can take place in four different ways.

These processes are conduction, convection, [11] _____ and evaporation. Three of these apply to all objects but [12] _____ needs the presence of a liquid.

Heat can pass through all materials by conduction. Conduction relies on the movement of particles; when particles become more energetic they transfer energy through collisions with neighbouring particles. The best conductors are [13] _____ and the worst are gases. Poor conductors are called [14] _____ .

Convection involves a flow of matter, so it only takes place in liquids and [15] _____ . Convection currents are caused by changes in [16] _____ . When the air around a central heating radiator is heated the air [17] _____ , becoming less dense. The warmed air rises above the surrounding colder, denser air.

All objects radiate energy in the form of [18] _____ _____ . These waves are in the [19] _____ region of the electromagnetic spectrum. The hotter the object, the more energy it radiates each second. Very hot objects also emit light and other electromagnetic waves.

Some objects are better than others at absorbing and emitting radiant energy. Dark colours are good absorbers and good [20] _____ of infrared radiation. Light colours are poor emitters and poor [21] _____ of infrared radiation. Silvered surfaces reflect infrared radiation in the same way that they reflect light.

Carbon dioxide and other greenhouse gases in the Earth's atmosphere also absorb infrared radiation; they then re-radiate it in all directions. This helps to maintain the Earth's temperature, but there is a danger that increasing levels of carbon dioxide are causing the Earth to become warmer.

Evaporation occurs at the [22] _____ of a liquid or moist object. The most energetic particles escape, taking [23] _____ with them, causing the remaining liquid to be cooled. The rate of evaporation from a liquid can be increased by increasing the surface area or the [24] _____ of the liquid and by removing vapour from the surrounding air.

Insulation

| conduction | convection | energy | reflect |

Keeping things warm in a cold environment requires [25] _____ . If a house is warmer than its surroundings, it loses most energy by conduction and convection. The diagram shows the energy flow through an uninsulated cavity wall. The energy flows through the brick by [26] _____ and through the air-filled cavity by [27] _____ .

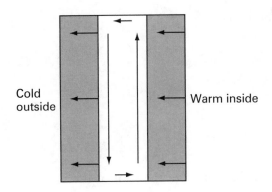

Cold outside

Warm inside

The rate of energy flow through the wall can be reduced by installing cavity wall insulation. This uses foam or mineral wool to trap pockets of air. If the air cannot move then [28] _____ currents cannot flow. Energy can still flow by the process of [29] _____ but gases are very poor conductors.

A hot drink or food taken from a hot oven is a lot warmer than its surroundings. It loses energy mainly by infrared radiation and evaporation. Aluminium foil is a very good insulator for hot food and drink because it [30] _____ infrared radiation and prevents hot vapour from escaping.

Summary questions for GCSE

31 Write down the **main** method of energy transfer for each example given in (a) to (d).
Choose from *conduction, convection, radiation* and *evaporation*.

(a) The energy flow through the bricks of a house wall.

.... [1]

(b) The energy lost from the surface of a hot drink.

.... [1]

(c) The energy given off by a red-hot grill.

.... [1]

(d) The energy transfer through an uninsulated cavity from the inner wall of a house to the outer wall.

.... [1]

[Total 4 marks]

32 The diagram shows the uninsulated loft space in a house.

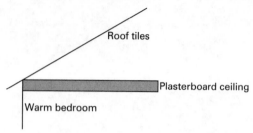

Roof tiles

Plasterboard ceiling

Warm bedroom

(a) Describe how energy from the warm bedroom is lost through the roof of the house.

....

....

.... [3]

(b) The Government recommends that a minimum thickness of 15 cm of insulation is laid on the floor of a loft. This insulation is usually in the form of glass fibres.

Explain how loft insulation reduces the energy loss from a house.

....

....

.... [3]

[Total 6 marks]

33 The energy flow through a window depends on the temperature difference between the inside and outside. The table shows the energy flow through a single glazed window for a range of temperature differences.

Temperature difference in °C	4	7	11	16	19	24
Energy flow per second in J/s	33	58	91	132	157	193

(a) Draw a graph of energy flow per second against temperature difference.

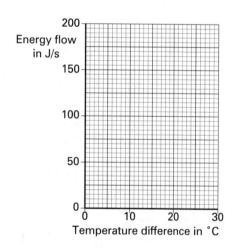

(b) Use your graph to find the energy flow per second when the temperature difference is

    (i)   5 °C .... J/s   [1]

    (ii) 10 °C .... J/s   [1]

    (iii) 20 °C .... J/s   [1]

(c) Is it true to say that 'doubling the temperature difference doubles the rate of energy flow through the window'? Use the data from (b) to justify your answer.

    ....

    .... [2]

(d) The energy flow through a double glazed window is approximately half of that through a single glazed window for the same temperature difference. Add a line to your graph that shows how the energy flow through a double glazed window depends on the temperature difference between the inside and the outside.

    Label your line 'double glazed'.   [2]

                                                    [Total 10 marks]

Radioactivity

Do all of this topic for Single Science and Double Science.
H = for Higher Tier only

 Revision notes

Types of radiation

alpha	aluminium	beta	electron
gamma	Geiger–Müller tube	lead	nucleus
photographic film			
electromagnetic radiation	*electron*	*helium nucleus*	

There are three main types of ionising radiation emitted when an unstable [1] _____

changes to a more stable form. These are called [2] _____ (α), [3] _____ (β)

and [4] _____ (γ). Complete this table, which compares the properties of these radiations.

ionising radiation	nature	symbol	penetration	detected by
alpha particle	two neutrons and two protons, referred to as a [H5] _____	$^4_2$He	absorbed by paper or a few cm of air	[6] _____ _____ and photographic film
beta particle	high-energy [H7] _____	$^0_{-1}$e	absorbed by thin sheets of [8] _____	Geiger–Müller tube and [9] _____
gamma ray	high frequency [H10] _____ _____		reduced by several cm of [11] _____	Geiger–Müller tube and photographic film

Using radioactivity

background radiation	beta	gamma	isotope	radiation
radioactive	*radiocarbon dating*			

All ionising [12] _____ is damaging to living cells. Low doses can cause temporary

sterility in both men and women; higher doses can burn the skin and eyes and very high doses damage the central nervous system. There is no safe way of handling [13] _____

materials, but the dangers are reduced if they are never handled directly and are kept at as great a

distance from the body as possible. Lead shields should be used where possible between the person

and the source of [14] _____ and the time of exposure should be kept to a minimum.

Carbon-14 is a form of carbon that is constantly being created in the atmosphere and its decay by

beta-emission contributes to the [15] _____ _____ , i.e. natural

radioactive emissions from the ground, the atmosphere and living things. When a plant or animal dies

it no longer takes in carbon-14 and so the level of carbon-14 present [16] *increases/decreases/stays*

the same as it decays. Measurements of the amount of carbon-14 present are used to determine the

age of once-living material in a technique known as [H17] _____

_____ . Rocks can also be dated using radioactivity; measurements of the relative

amounts of a [H18] _____ material and the product formed when it decays are used

to estimate the age of the rock.

Radioactive [19] _____ have a number of medical and non-medical uses. Alpha and

[20] _____ emitters are used to monitor the thickness of sheet materials. When used as

tracers [21] _____ emitters are preferred because their penetration allows them to be detected

easily. The fact that they are less ionising than alpha or [22] _____ emitters makes them

safer to use in the body. Radiotherapy also uses [23] _____ emitters to destroy cancers.

Radioactive decay

background radiation	decay	nucleus	random
half-life			

Radioactive decay is a [24] _____ process; the decay of an individual unstable

[25] _____ cannot be predicted. As the number of undecayed nuclei in a sample decreases,

so does the rate of [26] _____ . The average time it takes for half the unstable nuclei in a

sample to decay depends only on the substance and is known as the [H27] _____ .

After one [H28] _____ the activity of a radioactive substance, measured in counts/s

or becquerel, can be expected to halve. After two half-lives it is one [H29] *half/quarter/eighth* of the

original activity and so on. The graph shows how the activity of a radioactive material changes with

time, after corrections have been made by deducting the average count of the

[30] _____ _____ from each reading.

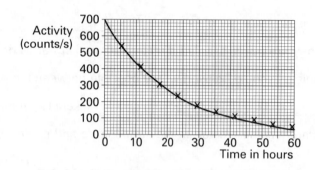

Summary questions for GCSE   H = for Higher Tier only

31 The three main types of radioactive emission are alpha, beta and gamma.

Which of these:

(a) is the most penetrative?

.... [1]

(b) is the most intensely ionising?

.... [1]

(c) are charged?

.... [1]

(d) can be detected by a Geiger–Müller tube?

.... [1]

[Total 4 marks]

32 If you fly in an aircraft you are exposed to a much higher level of background radiation than if you stand on Earth.

(a) Where does the higher level of background radiation come from?

.... [1]

(b) Suggest how a person standing on Earth is protected from this radiation.

.... [1]

(c) Which people are likely to be most affected by these higher levels of background radiation?

.... [1]

[Total 3 marks]

33 The table shows the results of an experiment to measure the half-life of radon-220, a radioactive gas formed when radium-224 decays. The readings have been corrected for the background radiation.

Time in s	0	20	40	60	80	100	120	140	160	180
Activity in counts/s	580	450	335	258	205	147	125	93	67	49

(a) Explain how readings can be corrected to take into account the background radiation.

....

....

.... [3]

(b) Draw a graph of activity against time. [3]

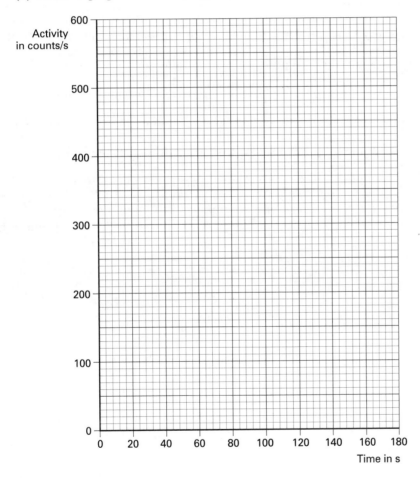

(c) Explain why all the points do not lie precisely on the line.

.... [1]

^H(d) Use your graph to calculate the half-life of radon.

....

....

.... [3]

[Total 10 marks]

34 Sodium-24 can be used to detect leaks in water pipelines. Sodium chloride containing sodium-24 is introduced into the water pipeline and a radiation detector is then used to check for leaks.

Sodium-24 has a half-life of 15 hours and it emits gamma radiation when it decays.

(a) What radiation detector could be used to detect leaks?

.... [1]

(b) Write down **three** reasons why sodium-24 is suitable for helping to detect leaks in water pipelines.

1 ....

2 ....

3 .... [3]

(c) What precautions should be taken by a worker whose job is to detect the leaks?

1 ....

2 .... [2]

[Total 6 marks]

The Solar System

Do all of this topic for Single Science and Double Science.
H = for Higher Tier only

 Revision notes

The Solar System

axis	electromagnetic radiation	galaxy	gravitational
light	light-year	Mercury	orbital
planet	Pluto	season	solar system
star	Sun	year	

The Sun is the centre of the [1] _____ _____ . The Sun is a small

[2] _____ in the Milky Way [3] _____ , one of millions of collections

of [4] _____ that make up the Universe.

It emits [5] _____ and other forms of [6] _____

_____ .

Astronomical distances are measured in [7] _____ , the distance that light travels

in one Earth [8] _____ .

The Earth and other [9] _____ are kept in orbit around the [10] _____ by the

[11] _____ attractive forces that exist between all masses. The greater the distance

from the Sun, the weaker its gravitational pull and so the more distant planets have lower

[12] _____ speeds than the inner planets.

The Earth makes a complete orbit of the Sun in one [13] _____ . Day and night are caused

by the Earth's spin on its own [14] _____ ; it completes one rotation in 24 hours. This spin

causes the [15] _____ to appear to rotate in the opposite direction.

Variations in the number of hours of daylight and the intensity of the Sun's radiation are the cause of

the Earth's [16] _____ ; they are due to the tilt of the Earth's axis.

The planet closest to the Sun is [17] _____ .

The planet furthest from the Sun is [18] _____ .

Satellites

Moon	orbit	satellite	Sun
asteroid	geosynchronous	Jupiter	Solar System

In a similar way to the motion of the planets around the [19] _____, the Earth's gravitational pull on the [20] _____ and artificial [21] _____ keeps them in orbit around the Earth.

Satellites in a low [22] _____ around the Earth have short orbit times; they are useful for monitoring changing weather patterns. The higher the satellite, the longer the [23] _____ time. A satellite placed above the equator with an orbit time of 24 hours is said to be [H24] _____ as it always stays above the same place on the Earth's surface.

The [H25] _____ belt lies between Mars and [H26] _____. Asteroids are fragments of rock thought to have been left over from the formation of the [H27] _____ _____.

Comets

comet	gravitational	Sun

The photograph shows a comet. Notice the tail on the comet, which becomes visible when the comet is close to the Sun. Comets orbit the [H28] _____ but unlike the planets, whose orbits are confined to a thin disc and which all travel the same way around the Sun, the orbits of [H29] _____ can be in any direction and any plane.

The shape of a comet's orbit is an ellipse. As it approaches the Sun it speeds up owing to the increasing [H30] _____ pull in the direction of travel.

As it moves away from the Sun, the gravitational pull on the comet becomes weaker; it follows an inverse square relationship which means that when the separation from the Sun is doubled the force is reduced to one quarter of its original size.

Summary questions for GCSE   H = for Higher Tier only

31 These objects are all part of the Universe.

galaxy   moon   planet   satellite   Sun

(a) Which **one** is a star?

.... [1]

(b) Which is the largest object in the Solar System?

.... [1]

(c) Which object orbits a Sun?

.... [1]

(d) Which **two** objects can orbit a planet?

.... [2]

(e) Which object is a collection of stars?

.... [1]

[Total 6 marks]

32 The table gives some data about the four inner planets and Jupiter.

Planet	Mass in kg	Radius in km	Time to rotate on its own axis in Earth days	Time to orbit the Sun in Earth years	Average distance from Sun compared with Earth
Earth	6.0×10^{24}	6380	1	1	1
Jupiter	1.9×10^{27}	71490	0.42	11.9	5.2
Mars	6.4×10^{23}	3400	1	1.9	1.5
Mercury	3.3×10^{23}	2440	59	0.24	0.4
Venus	4.9×10^{24}	6050	240	0.6	0.7

(a) Which **two** planets travel faster in their orbit than the Earth does?

.... [2]

(b) Which planet has the longest day?

.... [1]

(c) Write a list of the planets in order of their masses, starting with the least massive.

....

.... [2]

(d) Sketch the positions of the planetary orbits around the Sun. [2]

Sun

(e) Write down **two** reasons why it takes Mars nearly twice as long as it takes Earth to complete one orbit around the Sun.

....

.... [2]

(f) Draw a graph of time planets take to orbit the Sun against their distance from the Sun.

[3]

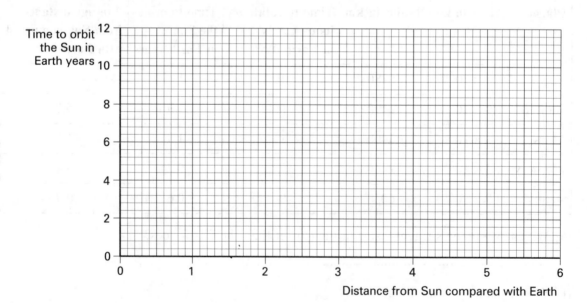

(g) What does the graph show about how planetary orbit times depend on the distance from the Sun?

....

.... [2]

(h) The asteroids orbit the Sun with orbit times between three and six Earth years.

Use the graph to find the range of distances of the asteroids' orbits.

.... [1]

(i) Between which **two** planets are the asteroids?

.... [1]

[Total 16 marks]

H33 (a) Explain how the orbits of comets differ from those of planets.

....

....

.... [3]

(b) Give **two** uses of artificial satellites.

....

.... [2]

(c) Explain **one** advantage of using geosynchronous satellites for communications.

....

....

.... [2]

[Total 7 marks]

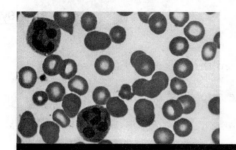

Cell mechanisms and circulation

Do all of this topic for Double Science only.
H = for Higher Tier only

module 6

 Revision notes *Vital exchanges*

Structure of plant cells

Label the drawing of the cell from a plant leaf.

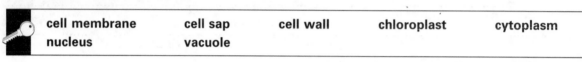

5 *cell wall*

6 *vacuole*

7 filled with *cell sap*

1 *chloroplast*

2 *cytoplasm*

3 *cell membrane*

4 *nucleus*

Functions of parts of plant cells

cellulose   chlorophyll   support

Plant cell walls are made from 8 *cellulose* for support. The chloroplasts contain
9 *chlorophyll* which is needed for photosynthesis. The vacuoles contain cell sap which
is used for storage and 10 *support*

Enzymes

carbon dioxide   catalyst   ethanol   lactic acid   optimum

Enzymes are biological 11 *catalyst* . The temperature at which an enzyme works
best is known as its 12 *optimum* temperature. Enzymes are used:

● in bread-making to produce 13 *carbon oxide*

● in brewing to produce 14 *ethanol*

● and in yoghurt-making to produce 15 *lactic acid* .

How substances enter and leave cells

> **cell membrane**
>
> **active transport**   **concentration gradient**   **diffusion**   **osmosis**
>
> **partially permeable**

To enter a cell a molecule or ion must pass through the [16] _cell_ _membrane_

The movement of a molecule or ion from a region of high concentration to a region of lower

concentration is called [H17] _diffusion_.

A [H18] _concentration gradient_ is the difference in concentration of a

substance in two areas.

Cell membranes that allow small molecules to pass through quickly, but large molecules more slowly,

are said to be [H19] _partially permeable_.

Diffusion of water molecules through a partially permeable membrane along

a [H20] _active transport_ is known as [H21] _osmosis_.

Movement of substances against a concentration gradient is known as [H22] _partially_

permeable.

Function of the circulatory system

> **back-flow**   **carbon dioxide**   **elastic**   **muscular**
>
> **oxygen**   **plasma**   **valve**

Checking the pulse is an important medical test. Each pulse is the result of one heart beat.

The heart is a double pump which forces blood around the body. It contains valves to prevent

the [23] _back-flow_ of blood.

Arteries have thick walls containing [24] _elastic_ tissue so that they can control the supply of

blood to the organs, and [25] _muscular_ tissue so that they can expand when the heart beats.

They contain blood at [26] _high/low_ pressure. Veins have thinner walls and, unlike arteries,

contain [27] _plasma_. Capillaries have walls one cell thick to allow [28] _plasma_

to flow out carrying food and oxygen to the tissues. The left side of the heart pumps blood that is rich

in [29] _oxygen_ into all arteries except the artery carrying blood to the lungs.

The right side of the heart receives blood rich in [30] _carbon dioxide_ from all

veins except the vein carrying blood from the lungs to the heart. This blood is pumped to the lungs to

release carbon dioxide and gain oxygen.

Summary questions for GCSE   H = for Higher Tier only

31 The drawing shows blood vessels associated with cells in the liver.

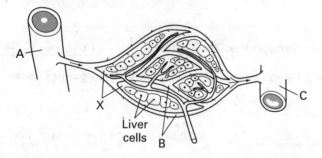

(a) Name the types of blood vessel labelled A, B and C. Give the reason for your choice in each case.

A   name: ...... vein .... [1]

reason: .... " .... runs .... back .... up ...to... the ......

.... heart. .... [1]

B   name: ...... capilleries .... [1]

reason: .... " .... run ... through ... into ....

.....small .... parts .... e.g. ... aveoli. .... [1]

C   name: ...... Artery .... [1]

reason: .... " .... runs .... to ... rest ... of ... body ......

.... [1]

(b) (i)   Name the fluid labelled X .... cellulose .... [1]

(ii) Where does this fluid come from?

.... liver ..... cells .... [1]

(c) Name **two** substances which diffuse from the blood plasma into the liver cells.

1 .... glucose .... [1]

2 .... [1]

(d) Name **two** substances which diffuse from the liver cells into the blood plasma.

1 .... urea .... [1]

2 .... carbon dioxide .... [1]

(e) When blood leaves the liver it passes to the heart and then to the lungs. Describe the path taken by the blood and the mechanisms which ensure that the blood reaches the lungs.

The de-oxygened blood flows through to the Artery and then the aorta while during this the valves are opening and closing to stop back-flow. The blood comes through to the heart on the left side and pumps through the left ventricle and out the semi-lunar valve into the lungs. The blood flows into the liver and gets cleaned and flows on to the next destination. [6]

[Total 18 marks]

H32 The drawing represents glucose molecules dissolved in water at a particular moment in time. Explain what would happen over the next few minutes.

Water

Glucose molecules

....

....

....

....

....

[Total 3 marks]

H33 Sultanas are dried grapes. If a sultana is soaked in water for a few hours it swells as shown in the drawings.

Sultana before soaking

Sultana after soaking

Explain why the sultana swells.

....

....

....

....

....

....

....

[Total 5 marks]

Breathing and respiration

Do all of this topic for Double Science only.
H = for Higher Tier only

Revision notes

Structure of the thorax

 alveolus   diaphragm   intercostal muscle   lung   rib
trachea

This diagram shows the organs inside the human thorax. Use key words to label these organs.

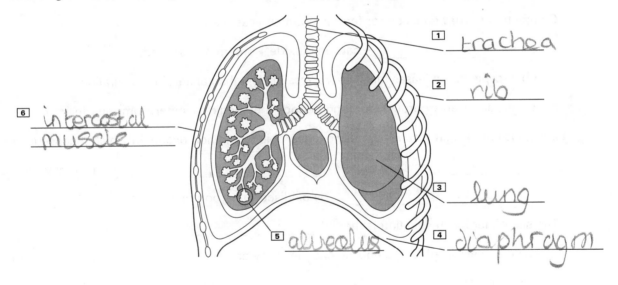

1	trachea
2	rib
3	lung
4	diaphragm
5	alveolus
6	intercostal muscle

Breathing

 carbon dioxide   intercostal muscle   oxygen   pressure   volume

The runner needs to breathe to take in the oxygen to release the energy he needs to run.

To breathe in, the [7] _intercostal muscle_ contract pulling the ribs upwards.

At the same time the diaphragm muscles contract pulling the diaphragm [8] *upwards*/~~downwards~~.

These movements increase the [9] _volume_ of the thorax, causing a decrease in

[10] _pressure_. Air then moves into the lungs. In the alveoli, [11] _oxygen_ diffuses from

the air into the blood and [12] _carbon dioxide_ diffuses from the blood into the air.

Respiration

aerobic	anaerobic	carbon dioxide	energy	ethanol	fatigue
glucose	lactic acid	oxidised	oxygen debt		
capillary	mitochondria	surface area			

Respiration is the process by which [13] _energy_ is released from food.

[14] _aerobic_ respiration uses oxygen, but [15] _anaerobic_ respiration does not use oxygen.

Complete the word equation for aerobic respiration:

[16] _glucose_ + oxygen → [17] _carbon dioxide_ + water + energy transferred.

The symbol equation for this reaction is $C_6H_{12}O_6 + 6O_2 \rightarrow 6CO_2 + 6H_2O$ + energy transferred.

Complete the word equation for anaerobic respiration in muscle cells:

[18] _glucose_ → [19] _lactic acid_ + energy transferred.

Complete the word equation for fermentation by yeast cells:

Glucose → [20] _ethanol_ + carbon dioxide + energy transferred.

Aerobic respiration releases far [21] more/~~less~~ energy than anaerobic respiration.

If muscles are used vigorously for long periods they begin to suffer from muscle [22] _fatigue_.

If insufficient oxygen reaches a muscle, anaerobic respiration occurs in which glucose is not

[23] _oxidised_, resulting in an [24] _oxygen debt_. To repay this debt,

[25] _____ _acid_ is oxidised to carbon dioxide and [26] _glucose_.

The site of aerobic respiration is the [H27] _capillary_.

Alveoli are specialised for gaseous exchange. They are [H28] moist/~~dry~~, thin-walled and folded to

increase their [H29] _surface area_ and are well supplied with

[H30] _mitochondria_

Summary questions for GCSE   H = for Higher Tier only

31 (a) Complete the sentences.

Air passes from the throat towards the lungs through the ...._bronchus_.... This branches to form ...._bronchi_.... which pass into each lung then divide to form many
...._alveoli_.............. Oxygen diffuses through the walls of the ...._alveolus_... into the blood capillaries.

[4]

(b) A student breathed out ten times into a bag. The volume breathed out was measured. The air was then analysed. The results are shown in the table.

number of breaths	10
total volume of air breathed out	6000 cm^3
volume of oxygen in air breathed out	1020 cm^3
volume of carbon dioxide in air breathed out	180 cm^3

   (i)   Calculate the mean volume of air breathed out in one breath.

$6000 + 180 = 6180$
$\dfrac{1020}{6180} = 16$

Answer ......600..... cm^3 [2]

   (ii)   Calculate the percentage of oxygen in the air breathed out.

$\dfrac{1020}{6000} = 0.17$

Answer ......17..... % [2]

(c) The student then exercised by doing step-ups for 2 minutes. She then breathed into the bag ten times.

   (i)   How would the volume of air breathed out after exercise compare with the volume breathed out before exercise? Explain the reason for your answer.

After excercise compared to before she would be breathing more deeply because the body is trying to gain more oxygen to distribute around of the body to keep it going. [3]

   (ii)   How would the percentage of carbon dioxide in the air breathed out after exercise compare with the percentage of carbon dioxide in the air breathed out before exercise? Explain the reason for your answer.

After excercise you would breath out more carbon dioxide than before because, the body needs to get rid of it quickly so it has take more oxygen. [3]

H(d) Explain **two** ways in which alveoli are adapted for efficient gaseous exchange.

1 ....

.... [2]

2 ....

.... [2]

[Total 18 marks]

H32 The drawing shows changes in pressure in the alveoli in one breathing cycle.

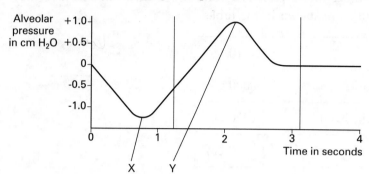

(a) Calculate the breathing rate.

Breathing rate .... breaths per minute [2]

(b) Explain the mechanisms which caused the pressure changes resulting in

(i) pressure **X**.

....

....

.... [3]

(ii) pressure **Y**.

....

....

.... [3]

[Total 8 marks]

H33 The table shows the results of two different exercises performed by the same athlete on successive days.

Exercise – rest periods	Total distance run in yards	Concentration of lactic acid in blood at end of exercise period in mg per 100 cm^3 of blood
4 minutes continuous	1422	150
10 s exercise – 5 s rest	7294	44

Explain why the blood lactic acid level was higher at the end of the 4 minutes continuous exercise even though the distance run was much shorter than in the intermittent exercise.

This is because, the 4 minute run
was continious and didn't *stop until finish
so it was working the body much harder than
the person who took brakes.

[Total 4 marks]

Plant physiology

Do all of this topic for Double Science only.
H = for Higher Tier only

Revision notes

The materials of photosynthesis

carbon dioxide	chlorophyll	chloroplast	glucose	light
oxygen	respiration	water		

Photosynthesis is the process which produces almost all the food we eat.

In order to photosynthesise, plants absorb [1] _carbon_ _dioxide_ from the air and

[2] _water_ from the soil.

[3] _light_ energy is absorbed by a green pigment called [4] _chlorophyll_

which is found in the [5] _chloroplast_ of leaf cells. The products of photosynthesis are a

sugar called [6] _glocuse_ and the gas [7] _oxygen_ . This gas can be used in

[8] _respiration_ by both plants and animals.

The sugars produced by photosynthesis

carbon dioxide	chlorophyll	energy	growth	oxygen
amino acid	*cellulose*	*lipid*	*starch*	

Most of the sugars are used up in respiration to release [9] _oxygen_ .

Some are converted into the many different molecules needed for the [10] _growth_ of young

cells, including [H11] _cellulose_ which is needed to produce cell walls.

Some are transported to other parts of the plant and stored as a carbohydrate called

[H12] _starch_ or as [H13] _energy_ in seeds. Others are combined with nitrates to

produce the [H14] _amino acid_ needed for growth and repair.

light [16] _____

[15] _carbon dioxide_ + water ⟶ glucose + [17] _oxygen_

[18] _starch_

How leaves are adapted for photosynthesis

| 🔑 | chloroplast | diffuse | phloem | stoma (plural stomata) | surface area | xylem |

Mid-rib
Vein

This diagram shows a leaf. Most leaves are wide and flat. This gives them a large

[19] _surface area_ to absorb light. The light energy is absorbed by the many

[20] _chloroplast_ inside the leaf cells. The leaves have tiny holes called

[21] _phloem_ on their under-surfaces to allow carbon dioxide to enter. Leaves are thin to

reduce the distance over which carbon dioxide has to [22] _diffuse_ . The veins of the leaf

transport materials. Water is transported to the leaves via a type of tissue called [23] _xylem_ .

Sugars are transported away from the leaves via a type of tissue called [24] _stoma_ .

Control of plant growth

| 🔑 | auxin | carbon dioxide | cutting | elongate | fruit |
| | root | temperature | weed | | |

Different factors limit the rate of photosynthesis at different times of the day and year. In a closed

greenhouse at noon on a summer's day the rate of photosynthesis is limited by low

[25] _carbon dioxide_ concentration. At noon on a sunny winter's day the rate of

photosynthesis is most limited by the low [26] _temperature_ .

Growth in plants is controlled and co-ordinated by [27] _auxin_ .

These auxins cause cells to [28] _root_ and prevent the growth of side branches.

Plant hormones can be used to kill [29] _weed_ in lawns or amongst crops.

A plant shoot that has been removed from the rest of the plant is known as a [30] _fruit_ . If

this is dipped in hormone powder it will form [31] _elongate_ . If hormones are applied to

flowers they will control the development of [32] _auxin_ .

Mineral requirements of plants

> chlorophyll   protein
>
> *active transport*

Root hair cells absorb mineral salts from the soil by [H33] active   transport .

Complete this table of the uses of mineral ions in plants and the effects of their deficiency:

mineral ion	function	effect of deficiency
nitrate	synthesis of [34] protein	stunted growth
iron and magnesium	synthesis of [35] chlorophyll	yellow leaves

Path taken by water through a plant

> evaporate   root hair   stoma   transpiration   xylem

Label this drawing showing the path taken by water through a plant:

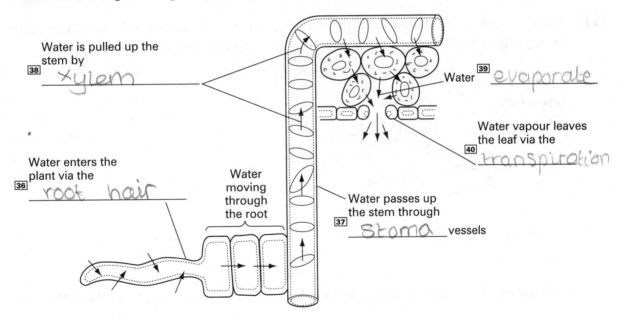

Water is pulled up the stem by [38] xylem

Water [39] evaporate

Water vapour leaves the leaf via the [40] transpiration

Water enters the plant via the [36] root hair

Water moving through the root

Water passes up the stem through [37] stoma vessels

Stomata

> guard cell   water   wilting

Stomata are holes between [41] guard   cell . If plants lose water faster than

they can absorb it the stomata may close to prevent the plant [42] wilting . The opening of stomata is caused by the entry of [43] _water_ into the [44] _guard_ _cell_ causing them to change shape.

Support

| cell wall | osmosis | plasmolysis | pressure | turgor |

As water moves into a plant cell by [H45] _osmosis_ it increases the [H46] _pressure_ on the [H47] _cell wall_ of the cell. This pressure is known as [H48] _turgor_ pressure and is the principal means of support for leaves and for young plants.

If the concentration of water outside a plant cell is less than that inside, water moves [H49] _in_/out of the cell, and the cytoplasm shrinks, losing contact with the cell wall. This is known as [H50] _plasmolysis_.

Summary Questions for GCSE   H = for Higher Tier only

51 Metal foil was wrapped round one leaf of a potted plant as shown in the drawing. The plant was left in bright light for 4 hours. The leaf was then removed and tested for glucose.

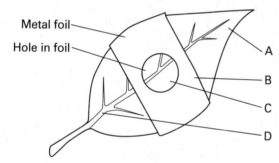

Metal foil
Hole in foil
A
B
C
D

(a) Which parts of the leaf, A, B, C and D, would give a positive result in the glucose test?

...A.. [1]

(b) Explain why some parts of the leaf would give a positive test for glucose but others would not.

....

....

.... [2]

(c) Give the word equation for photosynthesis.

...carbon dioxide + water → glucose + Oxygen + [3]
starch
[Total 6 marks]

52 A student set up the apparatus shown in the drawing. The lamp was set up at different distances from the pondweed. At each distance the student counted the number of bubbles given off in one minute. The experiment was repeated three times and the mean number of bubbles given off per minute was calculated.

The results are shown in the table.

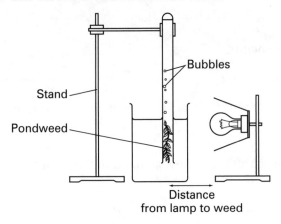

Distance from lamp to pondweed in cm	Mean number of bubbles given off per minute
5	61
10	20
15	9
20	4
25	3
30	3

(a) Describe the pattern shown by the results.

........The further the lamp is from the pondweed........ [2]
the lesser the bubbles.

(b) Explain the pattern shown by the results.

....

.... [3]

[Total 5 marks]

H53 Plants use normal carbon dioxide and radioactive carbon dioxide in exactly the same way. The leaf of a green plant was supplied with radioactive carbon dioxide as shown in the drawing. The plant was kept in the light. A few hours later starch in the root cells of the plant was found to contain radioactive carbon. Explain why this happened.

Supply of radioactive carbon dioxide

Clear glass container

....

....

....

....

....

....

....

[Total 6 marks]

54 A gardener planted out some seedlings early in the morning of a sunny day. The drawings show what had happened to the plants by noon on the same day.

Early morning   Noon

(a) Explain the change in the appearance of the plants.

....

....

.... [3]

(b) Explain why it would have been better for the gardener to have planted the seedlings in the early evening.

....

....

.... [3]

[Total 6 marks]

55

The drawing shows a potometer. It is used to compare the rate of water uptake by a plant shoot under different conditions. As the shoot takes in water the bubble moves towards the plant.

(a) Explain how the plant shoot takes in water from the beaker.

....

....

....

....

....

.... [3]

(b) A student placed the potometer on a bench in the laboratory and took five readings of the distance moved by the bubble in 10 minutes. The readings were 4.5 cm, 5.2 cm, 4.8 cm, 4.7 cm and 5.0 cm.

Calculate the mean distance travelled by the bubble in ten minutes.

Answer .... cm [2]

(c) The student then placed an electric fan near the potometer and again measured the mean distance travelled by the bubble in ten minutes.

How would you expect the result of this experiment to compare with that of the first experiment? Explain the reason for your answer.

....

.... [2]

(d) The student switched off the electric fan then placed a transparent plastic bag over the shoot.

How would you expect the result of this experiment to compare with that of the first experiment? Give the reason for your answer.

....

.... [2]

[Total 9 marks]

Energy flow and nutrient cycles

Do all of this topic for Double Science only.
H = for Higher Tier only

 ## Revision notes

Food chains

carnivore	consumer	energy	herbivore	omnivore
photosynthesis	producer			

This diagram shows food chains for some of the organisms in an aquarium.

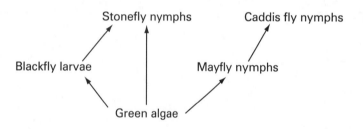

In the food chains, algae are [1] _____, mayfly nymphs are

[2] _____, caddis fly nymphs are [3] _____ and stonefly

nymphs are [4] _____ . Only the algae can carry out [5] _____ .

All the other organisms eat food and are therefore known as [6] _____ .

A food chain shows how [7] _____ is transferred between the organisms in a community.

The organisms whose food is obtained through the same number of links are said to belong to the

same trophic level.

Herbivores are also called primary consumers; carnivores are also called secondary or tertiary

consumers.

Ecological pyramids

The trophic levels of a community can be arranged in a pyramid shape, with producers at the base.

Pyramids of numbers illustrate the number of organisms in each trophic level. Pyramids of biomass

illustrate the total mass of biological materials at each trophic level.

These diagrams show four ecological pyramids: A, B, C and D.

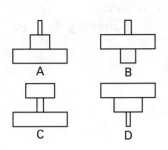

For the food chain: grass → rabbits → foxes, the pyramid of numbers is ⑧ *A/B/C/D* and the pyramid of biomass is ⑨ *A/B/C/D*.

For the food chain: oak trees → insects → birds, the pyramid of numbers is ⑩ *A/B/C/D* and the pyramid of biomass is ⑪ *A/B/C/D*.

The pyramid of numbers for the food chain: grass → cows → fleas is ⑫ *A/B/C/D*.

The carbon cycle

carbohydrate	combustion	fungi	microbe	photosynthesis
recycling	respiration			

When living organisms die their bodies may be decomposed by bacteria and by ⑬ _____ ,

so ⑭ _____ raw materials.

Green plants remove carbon dioxide from the atmosphere by ⑮ _____ and

convert it first into ⑯ _____ . Some of the carbon dioxide is returned to the

atmosphere by the ⑰ _____ of the green plants.

When green plants are eaten by an animal some of the carbohydrate becomes part of the animal.

Some of this carbohydrate is used by animals in ⑱ _____ to release energy;

carbon dioxide released by this process returns to the atmosphere.

When animals and plants die some animals and ⑲ _____ feed on their bodies. Carbon

dioxide is returned to the atmosphere during the ⑳ _____ of these organisms.

It is also returned to the atmosphere by the process of ㉑ _____ .

Energy loss in food chains

faeces	*heat*	*movement*

At each stage in a food chain less energy and materials are contained in the biomass of the organisms.

Some materials and energy are lost in the organism's H22 _____ . Animals need energy for

H23 _____ and much of this energy is lost to the environment as H24 _____ .

Eventually all the energy captured by green plants is returned to the environment.

The nitrogen cycle

| ammonium compounds | denitrifying bacteria | nitrates |
| nitrifying bacteria | nitrogen-fixing bacteria | putrefying bacteria |

Label this diagram of the nitrogen cycle:

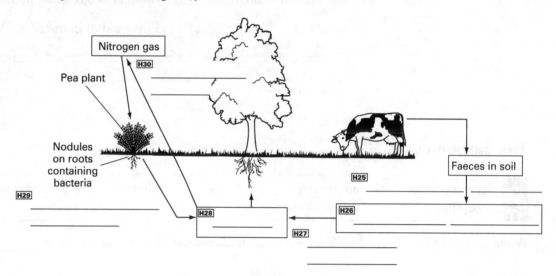

Nitrogen gas

H30

Pea plant

Nodules on roots containing bacteria

H29

H28

H27

H25

Faeces in soil

H26

Summary questions for GCSE

31 Some animals are useful to country gardeners because they eat animals that eat crops. Owls catch the mice which eat the gardener's sweetcorn. Shrews eat the caterpillars which eat the gardener's lettuces.

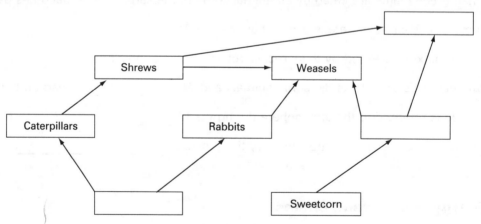

Shrews

Weasels

Caterpillars

Rabbits

Sweetcorn

(a) Use the information to complete the food web.   [3]

(b) From the food web:

  (i)   name **one** producer .... [1]

  (ii) name **one** herbivore .... [1]

  (iii) name **one** carnivore .... [1]

(c) (i)   Draw a pyramid of numbers for the food chain lettuces → caterpillars → shrews → owls.

[2]

(ii)   Draw a pyramid of biomass for the food chain sweetcorn → mice → weasels.   [2]

(d) When the gardener has harvested the sweetcorn, he puts the shoots of the plants onto a compost heap to decay.

(i)   Describe how living organisms are involved in the decay process.

....

....

....

.... [2]

(ii)   Under what **three** conditions will decay proceed fastest?

1 ....   2 ....   3 .... [3]

[Total 15 marks]

32   Label the processes **A**, **B**, **C** and **D** in the drawing of the carbon cycle.

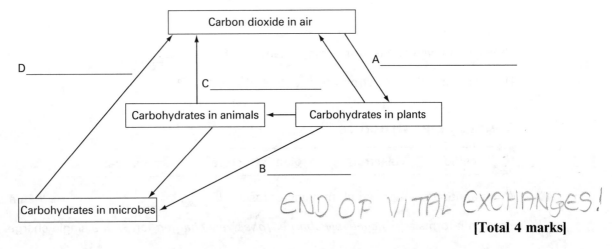

END OF VITAL EXCHANGES!

[Total 4 marks]

Atoms and bonding

Do all of this topic for Double Scienceonly.
H = for Higher Tier only

 Revision notes

Atoms and isotopes

atomic number	chlorine	electron	mass number	neutron
proton	sodium			
isotope				

Atoms are made up of three types of particles.

Use key words to complete this table:

particle	relative mass	charge
1 _____	1	0
2 _____	$\frac{1}{1840}$	−1
3 _____	1	+1

Atoms of the same element with different mass numbers are called H4 _____ . Chlorine-35

and chlorine-37 are two H5 _____ of chlorine. An atom of chlorine-35 contains 17 protons,

17 6 _____ and 18 neutrons. Chlorine-35 and chlorine-37 have the same

7 _____ _____ but different 8 _____ _____ .

The element with an electron arrangement of 2, 8, 1 is 9 _____ .

The element with an electron arrangement of 2, 8, 7 is 10 _____ .

Bonding and structure

| cubic | electron | giant structure | molecular |

When an ion is formed, an atom gains or loses 11 _____ . If an atom loses one

electron it forms a 12 *negatively charged/positively charged* ion with a single charge.

Substances can be divided into those which have [13] _____ structures and those

which have [14] _____ _____ . The photograph shows sodium chloride

crystals which are [15] _____ in shape. A regular shaped crystal is evidence for

[H16] *regular/irregular* arrangements of particles.

Substances with giant structures have [17] *high/low* melting points and [18] *high/low* boiling points.

Substances with molecular structures have [19] *high/low* melting points and [20] *high/low* boiling

points.

Ionic and covalent bonding

bonding	chlorine	ion	sodium	
covalent	*electrostatic*	*lattice*	*molecule*	*pair*

The forces which hold atoms together are called [21] _____ .

This diagram shows the arrangement of electrons in sodium and chlorine atoms.

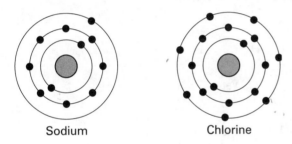

Sodium   Chlorine

When sodium and chlorine combine, one electron is transferred from the [22] _____ atom

to the [23] _____ atom. The resulting sodium and chloride [24] _____ are held

together in a structure or [H25] _____ by strong [H26] _____ forces.

This diagram shows the arrangement of electrons in one carbon atom and four hydrogen atoms.

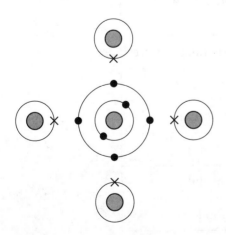

When the carbon atom and the four hydrogen atoms

combine they form a [H27] _____ of methane,

CH_4. Each bond uses one electron from the carbon atom

and one electron from a hydrogen atom to form a

shared [H28] _____ of electrons. Within each

methane molecule there are four [H29] _____

bonds. The forces between methane molecules are

[H30] *very strong/very weak.*

This diagram shows a methane molecule.

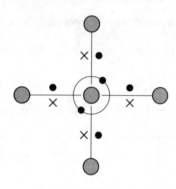

Summary questions for GCSE   H = for Higher Tier only

31 Hydrogen, carbon, oxygen and chlorine atoms have 1, 4, 6 and 7 outer electrons respectively. Finish the molecules showing the arrangement of outer electrons.

(a)
    Cl   Cl

    Chlorine

(b)
      O

   H   H

    Water

(c)
   H Cl

Hydrogen chloride

(d) O C O

Carbon dioxide

[Total 4 marks]

32 The table shows some properties of some elements and compounds.

Substance	Melting point in °C	Boiling point in °C	Electrical conductivity when:		
			solid	liquid	solution in water
A	370	872	poor	good	insoluble
B	420	908	good	good	insoluble
C	−112	−108	poor	poor	insoluble
D	1495	2877	good	good	insoluble
E	−70	57	poor	poor	good

(a) Which **two** letters represent metals?

.... and ....   [1]

Reason for your choice:

.... [1]

(b) Which **two** letters represent substances made up of molecules?

.... and ....   [1]

Reason for your choice:

.... [1]

(c) Which letter represents a substance which reacts with water to form ions?

....   [1]

Reason for your choice:

.... [1]

[Total 6 marks]

33 Carbon and silicon are elements in Group 4 of the Periodic Table. Both oxides contain covalent bonding.

(a) Suggest why carbon dioxide, CO_2, is a gas at room temperature.

....

....

.... [2]

(b) Suggest why silicon dioxide, SiO_2, is a solid at room temperature.

....

.... [2]

[Total 4 marks]

H34

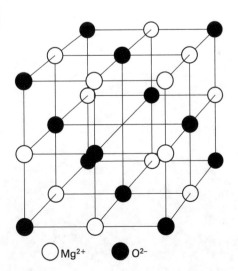

○ Mg^{2+} ● O^{2-}

The drawing shows a crystal of magnesium oxide. Magnesium oxide is formed when magnesium and oxygen combine.
The electron arrangements in magnesium and oxygen are: Mg   2, 8, 2
                                                          O   2, 6

(a) What structure does magnesium oxide have? .... [1]

(b) What type of bonding does magnesium oxide have? .... [1]

(c) Describe the changes in electron arrangement that occur when magnesium and oxygen combine.

....

.... [2]

(d) Sodium chloride and magnesium oxide have similar structure and bonding. The melting points of sodium chloride and magnesium chloride are 801 °C and 2730 °C respectively. Account for the difference in the melting points of these two compounds.

....

....

.... [2]

[Total 6 marks]

Quantitative chemistry

Do all of this topic for Double Science only.
H = for Higher Tier only

📝 Revision notes

The mole

🔑	atom	atomic	atmospheric	concentration
	formula mass	hydrogen	molar volume	mole
	molecule	volume		

The relative atomic mass of carbon is twelve. This means that the mass of one [1] _____ of

carbon is twelve times heavier than one atom of [2] _____ .

The amount of a substance containing 6×10^{23} particles is called 1 [H3] _____ . The

photograph shows 1 mole of carbon atoms.

The mass of 1 mole of a substance (called the [H4] _____ _____) can be found

by adding together relative [H5] _____ masses.

The mass of 1 mole of oxygen atoms is 16 g and the mass of 1 mole of oxygen

[H6] _____ is 32 g.

One mole of molecules of any gas occupies a [H7] _____ of 24 dm³ at room temperature

and [H8] _____ pressure. This is called the [H9] _____

_____ of a gas.

A solution of 1 mole of sulphuric acid dissolved in water to make 1000 cm³ of solution has a

[H10] _____ of 1 mol/dm³.

📄 Summary questions for GCSE   H = for Higher Tier only

11 | **Relative atomic masses**: H = 1, He = 4, C = 12, N = 14, O = 16, Al = 27, S = 32, Cl = 35.5, K = 39, Ca = 40, Fe = 56,
Cu = 64
The volume of 1 mole of any gas at room temperature and atmospheric pressure is 24 dm³.

(a) How many times heavier is one atom of copper than one atom of oxygen?

.... [1]

(b) How many atoms of helium have the same mass as one atom of calcium?

.... [1]

[H](c) There are 6×10^{23} atoms in 12 g of carbon.

What mass of chlorine contains the same number of atoms?

.... [1]

(d) What is the mass of 1 mole (formula mass) of sulphur dioxide, SO_2?

.... [1]

[H](e) A solution of nitric acid, HNO_3, has a concentration of 2 mol/dm^3.

What is the mass of 1 mole (formula mass) of nitric acid? ....

What mass of nitric acid is present in 100 cm^3 of this acid solution? .... g [2]

[H](f) What is the volume at room temperature and atmospheric pressure of:

(i) one mole of chlorine molecules? .... [1]

(ii) 28 g of carbon monoxide, CO? .... [1]

(iii) 11 g of carbon dioxide molecules, CO_2? .... [1]

[H](g) The masses of 1 mole of ethane gas and 1 mole of nitrogen monoxide are the same.

What does this tell you about the molecules in these gases?

.... [1]

[Total 10 marks]

[H12] The equation for the decomposition of calcium carbonate is:

$$CaCO_3(s) \rightarrow CaO(s) + CO_2(g)$$

The volume of 1 mole of any gas at room temperature and atmospheric pressure is 24.0 dm^3.

(a) Calculate the formula masses of calcium carbonate and calcium oxide.

Calcium carbonate: .... g   Calcium oxide: .... g   [2]

(b) How many moles of calcium carbonate are there in 500 g of calcium carbonate?

.... [1]

(c) How many moles of calcium oxide are formed when 500 g of calcium carbonate is decomposed?

.... [1]

(d) What volume of carbon dioxide is produced, at room temperature and atmospheric pressure, when 500 g of calcium carbonate is decomposed?

Answer ....dm^3 [2]

[Total 6 marks]

H13 A sample of dry copper oxide was reduced to copper using dry hydrogen. The apparatus used is shown in the diagram.

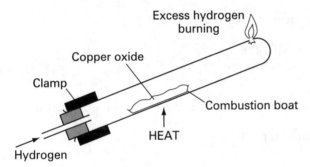

A known mass of the copper oxide was reduced until no further change in mass took place. The apparatus was then allowed to cool before the stream of hydrogen was stopped.
The results are shown below:

Mass of combustion boat = 10.62 g
Mass of combustion boat and contents before reduction = 14.22 g
Mass of combustion boat and contents after reduction = 13.82 g

(a) What mass of copper oxide was used? .... [1]

(b) What mass of copper was produced? .... [1]

(c) How many moles of copper atoms were produced?

.... [1]

(d) What mass of oxygen was removed from the copper oxide during reduction?

.... [1]

(e) How many moles of oxygen atoms were removed?

.... [1]

(f) What was the formula of the copper oxide?

.... [1]

(g) Why is it essential to allow the apparatus to cool before the hydrogen gas is stopped?

.... [1]

[Total 7 marks]

H14 When excess chlorine gas is passed over 9.0 g of aluminium, a reaction takes place to form 44.5 g aluminium chloride.

(a) Use the data to show that the simplest formula of aluminium chloride is $AlCl_3$.

[3]

(b) Complete and balance the symbol equation for the reaction.

.... $Al(s)$ + ... → .... $AlCl_3(s)$ [2]

[Total 5 marks]

H15 Potassium sulphate, K_2SO_4, is a compound which could be used as a potassium fertiliser.

(a) Calculate the formula mass of potassium sulphate.

.... [1]

(b) Calculate the mass of potassium sulphate needed to produce 2000 cm^3 of potassium sulphate solution of concentration 0.1 mol/dm^3.

.... [2]

(c) Calculate the percentage of potassium in potassium sulphate.

.... % [2]

(d) The equation for the formation of potassium sulphate from potassium hydroxide and sulphuric acid is:

$$2KOH(aq) + H_2SO_4(aq) \rightarrow K_2SO_4(aq) + 2H_2O(l)$$

(i)   How many moles of potassium hydroxide react with 1 mole of sulphuric acid?

.... [1]

(ii)   What mass of potassium hydroxide reacts with 1 mole of sulphuric acid?

.... [2]

(iii)   What mass of potassium sulphate would be produced by 5.6 g of potassium hydroxide?

.... [3]

[Total 11 marks]

Reversible reactions and transition metals

Do all of this topic for Double Science only.
H = for Higher Tier only

Revision notes

Reversible reactions

ammonia	equilibrium	product	reverse reaction	reversible

In some reactions all of the reactants are not turned into [1] _____ . As these form they react again to re-form the reactants. These reactions are called [2] _____ reactions.

If a reversible reaction is kept under constant conditions, an [3] _____ may be set up. Then the rate of the forward reaction is equal to the rate of the [4] _____ _____ . The concentrations of all the reactants and products remain unchanged unless the equilibrium is disturbed, when the equilibrium may move to the right (to produce more products) or to the left (to produce more reactants).

An example of a reversible reaction is:

nitrogen + hydrogen $\rightleftharpoons$ [5] _____

$$N_2(g) \ + \ 3H_2(g) \ \rightleftharpoons \ 2NH_3(g)$$

Decreasing the pressure moves the equilibrium to the [6] *left/right*, i.e. produces less [7] _____ . Removing ammonia from the equilibrium moves the equilibrium to the [8] *left/right* producing more ammonia.

A catalyst speeds up the forward reaction and the [9] _____ _____ . It does not produce more products but the equilibrium is established more quickly.

The Haber process and fertilisers

ammonia	exothermic	hydrogen	liquefying
nitric acid	nitrogen	oxygen	phosphorus
potassium	recycled	sulphuric acid	
catalyst	*eutrophication*	*iron*	*photosynthesis*

Ammonia gas is produced in the Haber process from three parts [10] _____ and one

part [11] _____ by volume. Fractional distillation of liquid air produces

[12] _____ and breaking down crude oil fractions or natural gas produces

[13] _____ . The gases are dried, mixed and compressed to a high pressure. They are then

passed over a [H14] _____ made of finely divided [H15] _____ heated to about

450 °C. About 10% of the gases are converted to ammonia. This is removed by

[16] _____ . The unreacted gases are [17] _____ .

The catalyst does not have to be heated during the process because the reaction is

[18] _____ .

The photograph shows a farmer spreading chemical fertiliser on the land. The fertiliser may contain

many chemical elements but three are essential for good plant growth. These are

[19] _____ (symbol N), [20] _____ (symbol P) and

[21] _____ (symbol K).

The essential element provided by ammonium nitrate, ammonium sulphate and urea is

[22] _____ . Ammonium nitrate is manufactured by the reaction of [23] _____ gas

with [24] _____ _____ . Ammonium sulphate is manufactured from ammonia

gas and [25] _____ _____ . Ammonium nitrate and ammonium sulphate

are very soluble in water and are therefore [26] *quick acting/slow acting*. Urea is almost insoluble in

water but reacts slowly to produce ammonia. It is therefore a [27] *quick acting/slow acting* fertiliser.

When nitrogen compounds are washed into rivers, they make the water plants grow well. The plants

prevent light entering the water and restrict [H28] _____ . When the plants die and

decay they use up [29] _____ in the water. The water becomes stagnant. This process is

called [H30] _____ .

Transition metals

| 🔑 | catalyst | copper | ion | manganese | transition metal | zinc |

The elements between Groups 2 and 3 in the Periodic Table are called

[31] _____ _____ . They are [32] *less reactive/more reactive*,

and [33] *denser/less dense* than alkali metals. They have [34] *high/low* melting points. Transition metals often form more than one metal 35 _____, e.g. Fe^{2+} and Fe^{3+}. Transition metal compounds are often coloured, for example iron(II) sulphate crystals are 36 *blue/green/brown/red*. Transition metals are good 37 _____ in industrial processes, e.g. iron in the Haber process. The transition metal used for electrical wiring is 38 _____ and for galvanising is 39 _____ . Another transition metal is 40 _____ .

Summary questions for GCSE   H = for Higher Tier only

H41 The Haber process is used to manufacture ammonia.

$N_2(g) + 3H_2(g) \rightleftharpoons 2NH_3(g)$

(a) The graph shows the percentage of ammonia produced at different temperatures and pressures.

(i)   What is the effect of increasing pressure on the equilibrium?
.... [1]

(ii) What is the effect of increasing temperature on the equilibrium?
.... [1]

(iii) The best conditions for making ammonia are 400 atmospheres and 350 °C.
      Why are these conditions **not** the ones used?

.... [1]

(b) An iron catalyst is used in the Haber process.

(i)   What effect does an iron catalyst have on the rates of the forward and reverse reactions?

....

.... [1]

(ii) What effect does an iron catalyst have on the yield of ammonia at equilibrium?

.... [1]

(iii) State **three** conditions necessary for a reversible reaction to establish 'dynamic equilibrium'.

1 .... [1]

2 .... [1]

3 .... [1]

(c) After passing the mixture of gases over the iron catalyst, the mixture contains ammonia, hydrogen and nitrogen. The boiling points of the three gases are given in the table.

Gas	Boiling point in °C
ammonia	−33
nitrogen	−253
hydrogen	−196

Use this information to explain how the ammonia is extracted from the mixture.

....

....

.... [3]

[Total 11 marks]

42 The table gives information about four fertilisers.

Name of compound	Formula	Solubility in water
ammonium nitrate	NH_4NO_3	very soluble
ammonium phosphate	$(NH_4)_3PO_4$	soluble
potassium nitrate	KNO_3	soluble
urea	$CO(NH_2)_2$	not very soluble in cold water but reacts slowly to produce ammonia and carbon dioxide.

(a) Write down the names of two fertilisers which will produce a 'complete' fertiliser on mixing.

.... [1]

(b) Write down a balanced symbol equation for the reaction of urea with water.

.... [2]

(c) Write an account of the problems caused to a stream when nitrogen compounds escape into it from the surrounding fields.

....

....

....

.... [4]

[Total 7 marks]

Electro-magnetism

Do all of this topic for Double Science only.
H = for Higher Tier only

 ## Revision notes

Magnetic poles and electromagnets

attract	current	electromagnet	magnetic field	north
pole	repel			

Magnets attract objects made out of magnetic materials such as iron, steel and nickel. They can

[1] a _____ and [2] r _____ other magnets. The strongest parts of a magnet are

called the [3] _____ . A fixed magnet has two poles, called the north and south poles. The

north (or north-seeking) pole of a magnet is attracted to the [4] _____ pole of the Earth and

the south (or south-seeking) pole is attracted to the Earth's south pole. Similar magnetic

poles [5] _____ each other and opposite poles [6] _____ .

Any region where a force acts on magnetic materials is called a [7] _____

_____ . Magnetic field patterns show the direction of the force on the [8] _____

pole of a magnet at each point in the field.

The most useful magnets are those that can be switched on and off; these are called

[9] _____ .

Every electric [10] _____ has its own magnetic field. A current passing in a coil of wire

creates a magnetic field both inside and around the coil. Using a [11] brass/plastic/soft iron core creates

a much stronger electromagnet. The core is quickly magnetised when the current passes in the coil

and it loses its magnetism quickly when the current is switched off.

Electric bell and relay

armature	attracted	contact screw	current	electromagnet
magnetic field				

The diagrams show two devices that use electromagnets, an electric bell and a relay.

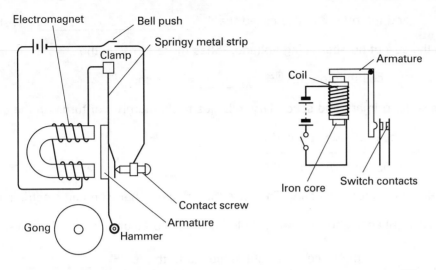

When the bell push is pressed, current passes in the coil of the [12] _____ .

The [13] _____ _____ of the coil magnetises the iron core and the

[14] _____ is attracted to the core, causing the hammer to strike the gong. This breaks the

circuit at the [15] _____ _____, causing the electromagnet to be switched off.

The [16] _____ springs back, making the circuit again so that the process is repeated.

The relay is a device that enables a small [17] _____ to switch a much larger current on and

off. When a small current passes in the coil, the magnetic field magnetises the iron core. This causes

the [18] _____ to be [19] _____ to the iron core, pressing the switch

contacts together.

Transformers and electric motors

alternating	current	magnetic field	voltage	
current	*magnetic field*	*primary*	*secondary*	*transformer*

Transformers use [20] _____ _____ to change the size of an

[21] _____ voltage.

Transformers are used in the electricity supply industry. To minimise the energy wasted as heat in the

transmission wires, electricity is distributed at a high voltage to keep the

[22] _____ as low as possible. Step-up transformers are used at power stations to

increase the [23] _____ before the electricity is fed into the grid. The voltage is stepped

down in stages before being supplied to homes and workplaces.

A [H24] _____ consists of two coils wound on an iron core. The input coil is called the [H25] _____ and the output coil is called the [H26] _____ . A step-up transformer increases the size of an alternating voltage; it has more turns on the [H27] _____ coil than on the [H28] _____ coil.

The equation that relates the primary and secondary voltages to the number of turns on the coils is:

$$\frac{voltage\ across\ coil\ 1}{voltage\ across\ coil\ 2} = \frac{number\ of\ turns\ on\ coil\ 1}{number\ of\ turns\ on\ coil\ 2}$$

Electric motors also rely on electromagnetism. They use the principle that when an electric current passes in a wire placed at right angles to a magnetic field, there is a force on the wire. The direction of the force is at right angles to both the [H29] _____ _____ and the [H30] _____, as the diagram shows.

In a motor, a coil of wire is placed between two opposite magnetic poles. Although the forces on the sides of the coil are in opposite directions, the moments about the coil axis are in the same direction.

Summary questions for GCSE   **H = for Higher Tier only**

31   Here are four different arrangements of two magnets.

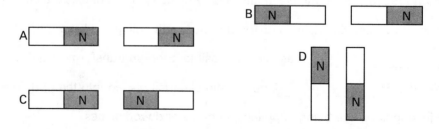

(a) Which show **two** magnets that are attracting each other?

.... [2]

(b) Which show **two** magnets that are repelling each other?

.... [2]

(c) Sketch the magnetic field pattern between the poles of the magnets in A.   [2]

[Total 6 marks]

32 Some houses have outside lights that switch on if a person approaches the house when it is dark. These lights have a low-voltage control circuit that operates a relay to switch on the mains lamp.

The diagram shows part of the circuit that is used.

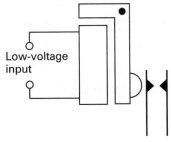

Low-voltage input

(a) Describe how the switch contacts become pressed together.

....

....

.... [3]

(b) Complete the diagram to show how the relay operates the mains lamp.   [2]

(c) Explain the advantage of using a relay in this device.

....

.... [2]

[Total 7 marks]

H33 The diagram shows the force acting on a wire placed between two opposite magnetic poles.

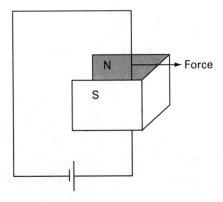

N

S

→ Force

(a) Label the directions of the current and the magnetic field on the diagram. [2]

(b) Write down the effect on the force of:

(i) swapping the magnetic poles.

.... [1]

(ii) reversing the current direction.

.... [1]

(c) Write down **two** ways in which the size of the force can be increased.

....

.... [2]

[Total 6 marks]

34 In a power station, electricity is generated at 25 000 V.

(a) The voltage is stepped up to 250 000 V before the electricity passes into the National Grid.

Describe how this is done.

....

.... [2]

(b) Explain why it is more economical to transmit electricity at a high voltage rather than a low voltage.

....

....

.... [3]

(c) Describe the process that takes place before electricity from the National Grid can be safely distributed to homes and workplaces.

....

....

.... [3]

(d) The electricity supply industry uses alternating current (a.c.). Discuss whether it would be possible to use direct current (d.c.) instead of alternating current.

....

....

.... [3]

[Total 11 marks]

Electrostatics and current

Do all of this topic for Double Science only.
H = for Higher Tier only

📝 Revision notes

Insulators and conductors

🔑	attract	conductor	electron	force	friction	negatively
	positively	repel				

An object can become charged by adding or removing [1] _____ from it. Charged

objects exert [2] _____ on each other; objects with similar charges. [3] _____

each other and those with opposite charges [4] _____ each other.

The [5] _____ forces that exist when two objects rub against each other cause the transfer

of electrons. The object that gains [6] _____ becomes

charged [7] _____ while the object that loses electrons becomes

charged [8] _____ . Insulators keep this charge

but [9] _____ quickly lose it as it passes through them and to earth.

Static charge

🔑	charge	conductor	electrostatic	spark	static	voltage

Charge that is not moving is said to be [10] _____ . It can be both useful and dangerous. A

build-up of static [11] _____ can create a high [12] _____ that can cause lightning

or sparks as it discharges, creating a fire hazard. When an aircraft is being

refuelled, [13] _____ are avoided by connecting the aircraft to earth using a good

electrical [14] _____ . Coal-burning power stations use charged plates

in [15] _____ smoke precipitators to remove dust from the waste gases.

Electric current

current	electron	negative	positive

A flow of charge is called an electric [16] _____ . Current in a metal is due to a flow of

[17] _____ that move away from the [18] _____ terminal of the battery

or power supply and towards the [19] _____ terminal. Current in conducting gases and

electrolytes is due to the movement of both positive and [20] _____ ions.

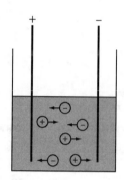

Movement of charged particles
in an electrolyte

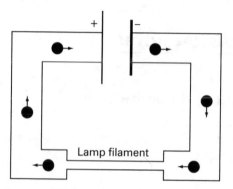

The movement of charge that makes current in a metal

Current and charge

energy	power	voltage	watt
charge	*coulomb*	*time*	*volt*

Electric current transfers [21] _____ . The rate of energy transfer is called the

[22] _____ . Power is measured in [23] _____ and is calculated using the

equation:

power = current × [24] _____ or $P = I \times V$

The unit of charge is the [H25] _____ (C); the amount of charge that flows when a current

passes is calculated using the equation:

charge = current × [H26] _____ or $Q = It$

The voltage between two points in a circuit is the energy transfer for each coulomb of

[H27] _____ that passes between the points.

$$voltage = \frac{energy\ transfer}{[H28] \underline{\hspace{3cm}}}$$

An energy transfer of 1 joule per [H29] _____ is a voltage of 1 [H30] _____ .

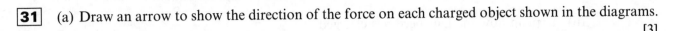

Summary questions for GCSE   H = for Higher Tier only

31 (a) Draw an arrow to show the direction of the force on each charged object shown in the diagrams.

[3]

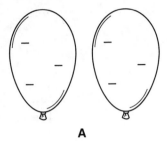 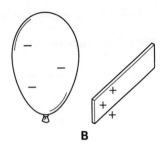 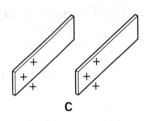

(b) In which diagram or diagrams is the force attractive?

.... [1]

(c) In which diagram or diagrams is the force repulsive?

.... [1]

[Total 5 marks]

32 When hair is combed using a nylon comb, some electrons move from the hair onto the comb.

(a) What type of charge does the comb gain?

.... [1]

(b) What type of charge does the hair gain?

.... [1]

(c) When the comb is held near the hair, some hair moves towards the comb.

Explain why this happens.

....

.... [2]

(d) When some false 'hair' is charged on a Van de Graaff generator, the hair rises away from the positively charged dome.

 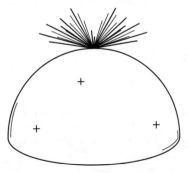

     (i)   What type of charge does the hair gain?

     .... [1]

     (ii) Explain how the hair becomes charged.

     .... [1]

     (iii) Explain why the hair rises from the dome.

     ....

     .... [2]

                                                                [Total 8 marks]

33   Calculate the power rating of each device shown in the table.

Device	Current in A	Voltage in V	Power in W
car headlamp	5	12	
microwave cooker	6	240	
iron	4.5	240	
soldering iron	1.5	12	
torch lamp	0.25	3	

                                                                [Total 5 marks]

H34   The current passing in a filament lamp is 0.25 A when it is operating from the 240 V mains.

     (a) Calculate the quantity of charge that passes through the filament in 60 s.

     ....

     ....

     .... [3]

     (b) How much energy is transferred to the lamp filament by 1 coulomb of charge?

     .... [1]

     (c) How much energy is transferred to the lamp filament in 60 s?

     .... [1]

     (d) Calculate the power of the lamp.

     ....

     ....

     .... [3]

                                                                  [Total 8 marks]

Force and energy

Do all of this topic for Double Science only.
H = for Higher Tier only

Revision notes

Work done and power

conserved	distance	force	joule
power	watt	work	

All energy transfers are due to forces doing [1]_____, making something move in the

direction of the [2]_____ . The work done by a force is calculated using the equation:

work done = force × [3]_____ *moved in the direction of the force.*

The work done is measured in [4]_____ (J) and is also equal to the energy transfer that

takes place. When a force does work energy is always [5]_____ ; any energy losses are

balanced by energy gains.

Power is a measure of how quickly work is done or energy is transferred; it is the rate of working or

transferring energy. The [6]_____ (W) is the unit of [7]_____ . Power is

calculated using the equation:

$$power = \frac{work\ done\ or\ energy\ transfer}{time\ taken} \quad or \quad P = \frac{E}{t} = \frac{Fd}{t}$$

Forces on an object

accelerating	balanced	braking distance	driving	force
friction	mass	reaction time	resistive	speed
stopping distance	thinking distance	unbalanced		
mass	*velocity*			

Any object moving at a constant velocity is not changing its [8]_____ or direction of

motion. The forces acting on it are [9]_____ . Changing speed or direction requires an

unbalanced or resultant [10]_____ . Forces always act between two objects; the forces they

exert on each other are [11] *equal/unequal* in size and act in [12] *the same/opposite* directions.

Motion on the Earth's surface is always opposed by resistive forces. The force that opposes slipping and sliding is called [13] _____ ; this is the force that stops your feet from sliding when you walk and prevents wheels from slipping when you ride a bike or travel in a bus, car or train. Air and water also exert [14] _____ forces that get bigger as you travel faster.

The friction force acting between the tyres of a vehicle and the road surface can also affect the [15] _____ _____ , which is the distance a vehicle travels from when the driver applies the brakes to when the vehicle stops. The braking distance also increases with increasing [16] s_____ and vehicle [17] m_____ . The total distance that a vehicle travels between a driver seeing a hazard and the vehicle stopping is called the [18] _____ _____ . This is made up of the braking distance and the [19] _____ _____ , which depends on the driver's [20] _____ _____ .

On the surface of the Earth, the Earth's gravitational field strength, g, has a value of 10 N/kg. The total weight of an object can be calculated using the equation:

weight = [21] _____ *× gravitational field strength* or *W = mg*

The forces acting on a car travelling at constant velocity on a level road are [22] _____ . To accelerate the car, the [23] _____ force must be greater than the [24] _____ forces. If the resistive forces are greater than the driving force the forces are [25] _____ and the car slows down.

An object that is changing its velocity is [26] _____ . Acceleration is calculated using the equation:

$$acceleration = \frac{change\ in\ [27]\ \rule{2cm}{0.4pt}}{time\ taken} \quad or \quad a = \frac{v - u}{t}$$

and is measured in m/s^2. The equation that relates the acceleration of an object to the size of the unbalanced, or resultant, force is:

force = [H28] _____ *× acceleration* or *F = ma*

Potential energy and kinetic energy

height   mass

Raising an object above the surface of the Earth increases its gravitational potential energy (gpe). The change in gpe when an object changes height is calculated using the equation:

gpe = weight × [H29] _____ or $E_p = W \times h = m \times g \times h$

Moving objects have kinetic energy, calculated using the equation:

kinetic energy = $\frac{1}{2}$ *×* [H30] _____ *× (speed)^2* or $E_k = \frac{1}{2} \times m \times v^2$

Summary questions for GCSE   H = for Higher Tier only

H31   Here is a graph showing the velocity of a car at various times during a journey.

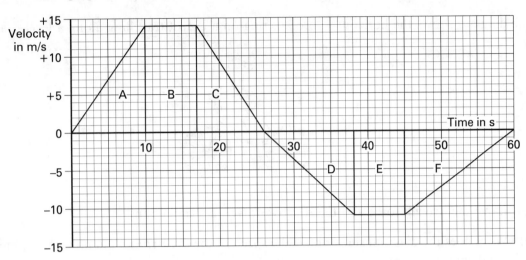

(a) Which two sections of the graph represent a decreasing speed?

.... [2]

(b) At what time shown on the graph does the car change direction?

.... [1]

(c) Calculate the acceleration of the car during the first 10 s.

....

....

.... [3]

H(d) The car and passengers have a total mass of 925 kg. Calculate the force required to cause the acceleration in the first 10 s.

....

....

.... [3]

H(e) The car is used to take a family on holiday. It carries the driver, three passengers and their luggage.

Explain how this affects the braking of the car. What advice would you give to the car driver?

....

....

.... [4]

[Total 13 marks]

32 (a) Write down **two** factors that could affect a cyclist's stopping distance.

1 ....

2 .... [2]

(b) The diagram shows the horizontal forces acting on a cyclist travelling along a level road.

(i) Label the driving force and the resistive force on the diagram. [2]

(ii) What does the diagram show about the speed of the cyclist? Explain how you can tell from the diagram.

....

.... [2]

(iii) Use the next diagram to draw and label the forces acting on a cyclist who is accelerating.

[2]

[Total 8 marks]

H33 A ball has a mass of 0.02 kg. It is thrown vertically upwards with a speed of 8.5 m/s.

(a) Describe the energy transfer that takes place as the ball rises and falls.

....

.... [3]

(b) Calculate the kinetic energy of the ball as it is released.

....

.... [3]

(c) Write down the increase in the gravitational potential energy of the ball at its maximum height. What does this assume about the forces acting on the ball?

.... [2]

(d) Calculate the maximum height that the ball reaches above the point of release.

....

.... [3]

[Total 11 marks]

Pressure and sound

Do all of this topic for Double Science only.
H = for Higher Tier only

 Revision notes

Pressure

force	hydraulic	pressure	proportional	random
pressure	*proportional*			

The effect that a force has in cutting or piercing is called the [1] _____ . The pressure

depends on the area that a force acts over and is calculated using the equation:

$$pressure = \frac{[2] \underline{\hspace{3cm}}}{area}$$

Skis have a large area so that the skier's weight causes a [3] *small/large* [4] _____ on the

snow and the skis do not penetrate it. Drawing pins have tips with a small area, creating a

[5] *small/large* pressure when they are pushed so that they can pierce materials.

If you pull on a bicycle brake lever, the [6] _____ is transmitted to the brake through the

cable. Liquids transmit the [7] _____ that is applied to them. This gives them the ability to

multiply [8] _____ and makes them useful in brakes for motor vehicles and for

[9] _____ machinery that is used for lifting cars, moving earth and shaping steel

panels in presses. The photograph shows hydraulics applied in an excavator.

Small force

Larger force

The diagram shows the principle of a hydraulic system to multiply [10] _____ . A force is

applied to the small piston, causing a [11] _____ to be transmitted through the liquid.

This pressure acts in all directions so it pushes on the large piston, causing a force that is

12 _____ to the area of the piston.

At any point in a liquid or gas, the pressure has the same value in all directions. The pressure is greater at a greater depth.

When a gas is subjected to an increasing pressure, it squashes.

The particles in a gas have large spaces between them and the motion of each particle cannot be predicted; it is 13 _____ both in speed and direction. Pressure is a result of collisions between the particles and the container walls.

The diagram illustrates the structure of a gas and the effect of squashing the gas into a smaller volume.

When the gas is squashed, collisions between the particles and container walls are more frequent, causing the H14 _____ on the walls to increase.

Provided that the temperature does not change, when the pressure on a gas is doubled the volume H15 *halves/doubles*. The pressure and volume are said to be inversely

H16 _____ .

Sound

amplitude	echo	frequency	light	longitudinal
reflection	speed	ultrasound	vacuum	vibrating
diffracted				

Sound is a 17 _____ wave that travels as a series of squashes and stretches due

to the particles of a material 18 _____ . The pitch of a sound is determined by

the 19 _____ and the loudness depends on the 20 _____ of

the vibration. Unlike 21 _____ and other electromagnetic waves, sound cannot be

transmitted through a 22 _____ .

Reflections of sound are called 23 _____ . Echo-sounding is a method of measuring

distance that uses 24 _____ , which has a frequency above the range of human hearing. A pulse of ultrasound waves is emitted and the time for the [25] _____ to return is measured. The **total** distance travelled by the ultrasound pulse can be calculated using the equation:

distance = [26] _____ × *time.*

Reflections of [27] _____ are also useful for scanning body organs and the fetus of a pregnant woman. The [28] _____ occur at tissue boundaries, enabling a computer to build up a picture of the organ or fetus.

Sound is refracted when its [29] _____ changes and it is readily

[H30] _____ , spreading out as it passes through windows and doorways which are of a similar size to the wavelength of the sound.

Summary questions for GCSE   H = for Higher Tier only

31   The diagram shows a hydraulic lifting mechanism.

Area of small piston = 0.25 m²   Area of large piston = 3.6 m²

200 N

Hydraulic fluid

(a) Calculate the pressure that the small piston exerts on the fluid.

....

....

.... [3]

(b) Write down the value of the pressure that the fluid exerts on the large piston.

.... [1]

(c) Calculate the size of the force that the fluid exerts on the large piston.

....

....

.... [3]

(d) Describe **two** advantages of using hydraulics in machines.

....

.... [2]

[Total 9 marks]

H32 A cylinder of carbon dioxide contains gas at high pressure.

(a) Explain how the gas exerts a pressure on the cylinder.

....

.... [2]

(b) The cylinder valve is opened and some gas escapes.

Explain how this affects the pressure of the gas left in the cylinder.

....

.... [2]

[Total 4 marks]

33 Two people are listening to a sound from a loudspeaker.

(a) Choose words from *smaller/greater/the same* to compare the properties of waves detected at **B** with those detected at **A**.

(i) the wavelength

.... [1]

(ii) the frequency

.... [1]

(iii) the amplitude

.... [1]

(b) The sound is now made to be louder.

What changes take place to the sound waves?

....

.... [2]

(c) Explain how the sound wave changes when a lower-pitched sound is produced.

....

.... [2]

[Total 7 marks]

H34 The diagram represents sound waves approaching a doorway.

(a) Draw in the wavefronts after the sound has passed through the doorway.   [2]

(b) Write down the name of this effect.

.... [1]

(c) Explain how the behaviour of a beam of light waves passing through a doorway is different to that of sound waves.

....

.... [2]

[Total 5 marks]

The Earth, its atmosphere and waves

Do all of this topic for Double Science only.
H = for Higher Tier only

Revision notes

The Earth's atmosphere

burned	global warming	greenhouse effect	mixture	oxygen
photosynthesis	respire			

Composition of the Earth's atmosphere by volume

Nitrogen	78.0%
Oxygen	20%
Argon	1%
Carbon dioxide	0.03%

together with water vapour in varying amounts and other noble gases in very small amounts

The air is a [1] _____ of gases. The active gas in the air is [2] _____ . The composition of air remains approximately constant because some processes use up oxygen and produce carbon dioxide while other processes use up carbon dioxide and produce oxygen. Oxygen is used up and carbon dioxide is produced when plants and animals [3] _____ or when fuels containing carbon are [4] _____ . Green plants remove carbon dioxide from the air and replace it with oxygen in the process of [5] _____ .

Over the last century the levels of carbon dioxide in the atmosphere have [6] *decreased/increased/ remained the same* because of the increasing amounts of carbon-rich fuels being burned and the destruction of forests. The build-up in levels of carbon dioxide is increasing the

[7] _____ _____ and may lead to [8] _____

_____ .

Origin of the Universe

	Big Bang	galaxy	Universe

Evidence for the origin of the Universe comes from the relative movement of the

[H9] _____ away from each other. The speed of movement is measured by the amount that

the colour of the light emitted by a galaxy is shifted towards the red end of the spectrum when it is

detected. The more distant galaxies appear to be moving away faster than the closer ones.

Measurements of the speeds of the galaxies show that they could have started at a single point.

Together with the microwave energy that fills space, thought to be radiation left over from an

explosion, this supports the [H10] _____ _____ theory of the way in which the

Universe began. According to this theory, the [H11] _____ started with a huge explosion

and the galaxies have been moving away from each other since then.

Waves

critical angle sound	energy total internal reflection	light transverse	longitudinal vibration
core *seismometer*	*frequency* *transverse*	*P* *wavelength*	*S*

Waves can transfer [12] _____ between places without the need for any material to move

between those places.

Waves are classified as transverse or [13] _____ . Where the

[14] _____ are parallel to the direction of wave travel, as in [15] _____

and underwater compression waves, the wave is [16] _____ . Vibrations at right

angles to the direction of wave travel, as is the case with [17] _____ , form

[18] _____ waves.

For all waves, the speed is related to the wavelength and frequency by the equation:

speed = [H19] _____ × *wavelength*   or $v = f \times \lambda$

Reflection of light takes place at the internal surface of glass and perspex. If the angle of incidence is

greater than the [20] _____ _____ , all the light is reflected when it hits the

boundary. This is called [21] _____ _____ _____ and is

used in reflecting prisms and optical fibres that transmit information.

Diffraction is a property of all waves. It occurs when waves pass through a narrow gap or past the edge of an obstacle. Diffraction is most noticeable when the size of the gap is equal to the H22 _____ of the waves. Very little spreading occurs at a gap that is many wavelengths wide.

Both longitudinal and H23 _____ waves are emitted when an earthquake occurs. The longitudinal waves are called P and the transverse waves are called S waves.

H24 _____ waves can travel through liquids and solids but H25 _____ waves can only travel through solids.

Both P waves and S waves can be detected by H26 _____ close to the centre of an earthquake, but only H27 _____ waves can be detected in the shadow region on the opposite side of the Earth. This gives evidence that the H28 _____ of the Earth is solid, since it allows both types of wave to pass through, but part of the H29 _____ is liquid, allowing only H30 _____ waves to pass through it.

Summary questions for GCSE   H = for Higher Tier only

31 (a) Which is the most plentiful **compound** in the Earth's atmosphere?

.... [1]

(b) The atmosphere contains other gases such as sulphur dioxide and oxides of nitrogen.

What are the main sources of these gases?

....

.... [2]

[Total 3 marks]

H32 The table gives some information about the composition of the Earth's atmosphere in the past.

Write an account of the processes which led to the formation of the Earth's atmosphere we have today.

Millions of years ago	Gases in the atmosphere
4500–4000	CO_2, steam, N_2, CH_4, NH_3, H_2S
2000	O_2 in small concentrations
1000	ozone starts to form
400	similar to today

....

....

....

....

....

....

[Total 6 marks]

33 The diagram shows how light travels in a curved optical fibre.

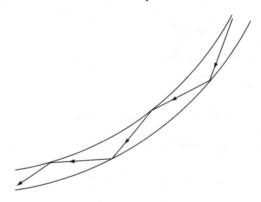

(a) Explain why light does not pass through the edges of the fibre.

....

.... [2]

(b) Describe how a surgeon can use optical fibres to examine the windpipe of a patient.

....

....

.... [3]

(c) Prisms can be used to turn the path of light through 90° or 180°.

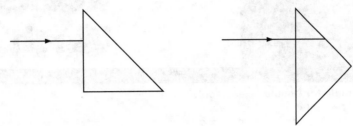

(i) Complete the diagrams to show the path of the light through the prisms.   [4]

(ii) Write down one use of a reflecting prism.

.... [1]

[Total 10 marks]

H34 Our Sun is thought to have formed in a dust cloud that was left over from an exploding supernova.

(a) Describe how the Sun is likely to have formed.

....

....

.... [3]

(b) What is the source of the energy that radiates from the Sun?

....

.... [2]

(c) Outline the probable future of the Sun.

....

....

.... [3]

[Total 8 marks]

Rocks in the Earth

Do all of this topic for Double Science only.
H = for Higher Tier only

 Revision notes

Types of rocks

calcium carbonate	crystal	extrusive	fossil	igneous
intrusion	intrusive	magma	marble	metamorphic
radioactivity	sedimentary			

Rocks can be divided into three types. Rocks which are formed when the magma cools and crystallises are called [1] _____ rocks. Two examples are granite and basalt. Granite is made up of large [2] _____ formed when the magma cools [3] *quickly/slowly* inside the Earth. A rock which crystallises inside the Earth rather than outside is called an

[4] _____ rock. Basalt is made up of small [5] _____ formed when the magma cools [6] *quickly/slowly* on the surface of the Earth. Rocks such as basalt, formed on the surface of the Earth, are called [7] _____ rocks.

Rocks which are formed when sediments are deposited and compressed are called

[8] _____ rocks. The photograph shows a chalk cliff. Different layers can be clearly seen. Rocks in the lower layers are [9] *younger/older* than the rocks above them.

When sedimentary or igneous rocks are subjected to [10] *high/low* pressures and [11] *high/low* temperatures, they may change and form [12] _____ rocks. In Italy, for example, limestone has been turned into the metamorphic rock [13] _____ which is used to make statues or for facing buildings. Both limestone and marble are forms of the chemical compound

[14] _____   _____ .

The diagram shows the rock cycle. Label the diagram using key words.

Water vapour
of volcanic origin

Weathering
and erosion

Sediments transported

Sea

Slow
uplift to
surface

Sediments deposited

15
_____ rocks
e.g. basalt

16
_____ rocks
e.g. granite

Crystallisation

19
_____ rocks
e.g. sandstone

Pressure

Heating
and
pressure

18
_____ rocks
e.g. slate

17
_____

Melting

The rock cycle

The structure in the diagram below is called an ⑳ _____ .

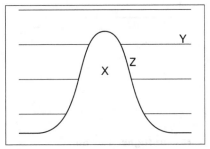

Rocks at point **X** are ㉑ _____ rocks.

Rocks at point **Y** are ㉒ _____ rocks.

Rocks at point **Z** are ㉓ _____ rocks.

The presence of ㉔ _____ can be used to date rocks. Another way of dating rocks is to

make ㉕ _____ measurements.

Structure of the Earth

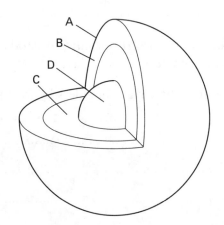

boundary	convection	core	crust
fault	mantle	plate	slide
constructive	*continental*	*destructive*	*magma*
oceanic	*subduction*		

The diagram shows the Earth's structure. The layer labelled **A** is the very thin layer of surface rock called the 26 _____ . The layer labelled **B** lies beneath the crust and is called the 27 _____ . There is movement by 28 _____ in the mantle. The labels **C** and **D** represent the 29 _____ .

The outer surface of the Earth is split into seven rigid sections or 30 _____ , each about 100 km thick. They are moving very slowly.

Plates can move in three different ways. An earthquake occurs when two plates 31 _____ past each other. Stresses build up along cracks or 32 _____ . These stresses are released when an earthquake occurs. There are a few earthquakes at the centres of plates, but most earthquakes occur at plate 33 _____ .

When two plates move apart, a H34 _____ plate margin exists. New rocks are formed as H35 _____ comes to the surface.

When two plates move together, a H36 _____ plate margin occurs. Here the weaker H37 _____ plate is forced underneath the H38 _____ plate, returning rocks to the H39 _____ . This is called H40 _____ .

Summary questions for GCSE   H = for Higher Tier only

41 (a) Sugar Loaf Mountain in Brazil is a granite mountain overlooking Rio de Janeiro. The granite was formed inside the chimney of a volcano.

How did the mountain become exposed?

.... [1]

(b) The Devil's Marbles are a group of large granite boulders in the desert of central Australia. The climate in the desert is very dry with large variations in temperature every day.

Suggest how the original granite has been broken down into these giant boulders by weathering in central Australia.

....

....

.... [4]

(c) Great Staple Tor is a group of granite boulders on Dartmoor in south-west England.

The climate on Dartmoor is cool and wet, with frost and snow in winter.

Suggest how these weather conditions have helped to break down the mass of granite into giant boulders on Dartmoor.

....

....

.... [4]

[Total 9 marks]

42 The diagram shows a cross-section through layers of different rocks in the Earth's crust.

(a) Write down **two** things which indicate that rocks are sedimentary rocks.

1 .... [1]

2 .... [1]

(b) Mark a letter **M** on the diagram in the place where marble might be found. [1]

(c) On the diagram, **X–Y** is a fault and **R–S** is a fold.

How can you tell that the fault **X–Y** occurred earlier in geological time than fold **R–S**?

....

.... [2]

[Total 5 marks]

H43 The map shows the distribution of plates on the surface of the Earth.

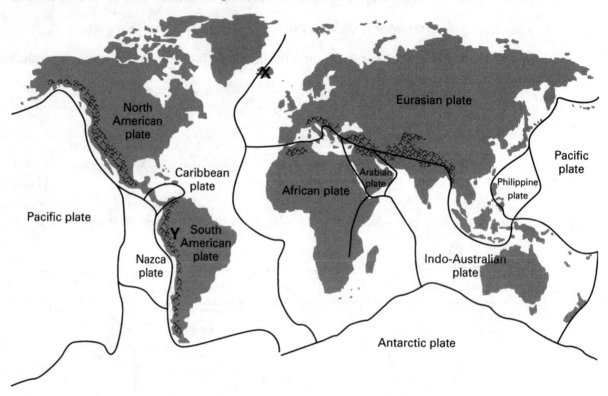

Use the map to help you answer these questions.

(a) Explain why the island of Iceland (labelled **X** on the map) is made up almost entirely of igneous rock.

....

....

.... [3]

(b) Explain how the Andes mountains labelled **Y** on the map were formed.

....

....

.... [3]

[Total 6 marks]

Using rocks and ores

Do all of this topic for Double Science only.
H = for Higher Tier only

 ## Revision notes

Extraction of metals

bauxite	electrolysis	haematite	open cast
reduction	rock	uncombined	

The photograph shows ore being dug out of the ground. This type of mining is called

[1] _____ _____ mining. An ore is a [2] _____ containing a metal or

compounds of a metal.

A common ore of iron is called [3] _____ and a common ore of aluminium is

called [4] _____ .

Metals high in the reactivity series, such as sodium and aluminium, are extracted from their ores by

[5] _____ . Metals in the middle of the reactivity series, such as zinc and iron, are

extracted by [6] _____ . Metals low in the reactivity series, such as gold, may be

found [7] _____ in the Earth.

Extraction of iron

air	blast	carbon	carbon dioxide
coke	reduced	reducing agent	

Iron is extracted from iron ore in a [8] _____ furnace. Iron ore, [9] _____ and

limestone are loaded into the furnace. Blasts of hot [10] _____ are blown into the furnace.

Coke is a form of the element called [11] _____ . Coke burns to form [12] _____

_____ which is reduced by more coke to form carbon monoxide.

Iron oxide in the ore is [13] _____ to iron by carbon monoxide. Carbon monoxide is the

[14] _____ _____ .

Extraction of aluminium

aluminium	anode	burn	carbon
carbon dioxide	cathode	electolysis	oxygen
cryolite			

Aluminium is extracted from purified aluminium oxide by [15] _____ .

Pure aluminium oxide is dissolved in molten [H16] _____ . The electrodes are made of

[17] _____ . An electric current is passed through the electrolyte. The product at the negative

electrode (or [18] _____) is [19] _____ . The element produced at the

positive electrode (or [20] _____) is [21] _____ .

The positive electrodes have to be replaced often because they [22] _____ to produce

[23] _____ _____ .

Purification of copper

| anode | cathode | electrolysis | electrolyte | ion |

The diagram shows the purification of impure copper by [24] _____ .

Use key words to finish the labelling.

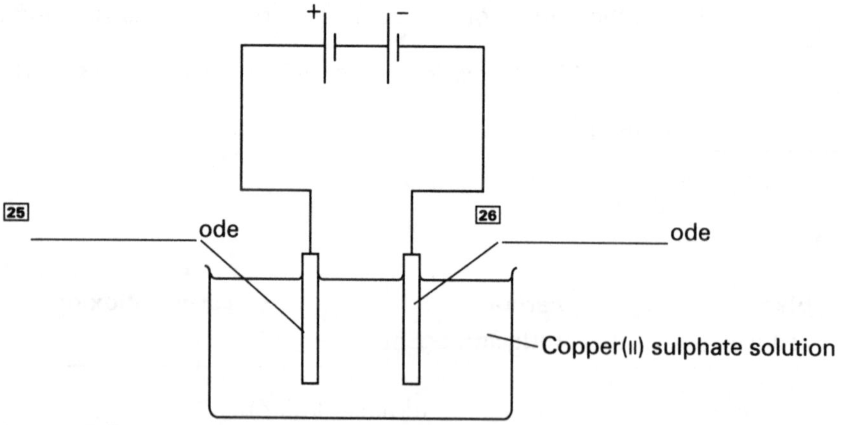

[25] _____ ode   [26] _____ ode

Copper(II) sulphate solution

Copper(II) sulphate solution in this process is called the [27] _____ .

Copper is produced at the [28] _____ .

Copper at the [29] _____ dissolves in the copper(II) sulphate solution to replace the copper

deposited.

Copper(II) sulphate solution contains positive and negative [30] _____ .

Uses of metals

| alloy | electricity | expensive | steel |

Most of the iron produced in the blast furnace is turned into **31** _____ which is an alloy of iron with carbon. Pure aluminium is used for overhead power cables because it is a good conductor of **32** _____ . Aluminium **33** _____ is used for aircraft panels because it has a lower density than most metals and is stronger than pure aluminium.

Aluminium is an expensive metal, although its ores are rich in aluminium, because of the cost of the **34** _____ required to make it. Any metal which has many uses and is difficult to extract from its ores is likely to be **35** _____ .

Summary questions for GCSE   H = for Higher Tier only

The reactivity series (on page 38) may be useful in answering these questions.

36 Zinc is found in the Earth in the ore called zinc blende. Zinc blende contains zinc sulphide, ZnS. Here are some statements about the way zinc is made from zinc blende.
They are in the wrong order.

A Zinc oxide is formed.
B Zinc vapour and carbon monoxide are formed.
C The zinc vapour is cooled to give solid zinc.
D Zinc blende is heated in air.
E Zinc oxide is mixed with carbon and heated.

(a) Put the statements in the correct order. Two have been done for you.

D → .... → .... → .... → **C**   [2]

(b) Finish the word equation for the reaction

zinc oxide + ... → zinc + ...   [2]

(c) Put a ring round an element which could be used instead of carbon to make zinc.

copper   hydrogen   iron   magnesium   [1]
[Total 5 marks]

37 The main equations in the extraction of iron are shown in the box below.

A	$C + O_2 \rightarrow CO_2$
B	$CO_2 + C \rightarrow 2CO$
C	$Fe_2O_3 + 3CO \rightarrow 2Fe + 3CO_2$
D	$CaCO_3 \rightarrow CaO + CO_2$
E	$CaO + SiO_2 \rightarrow CaSiO_3$

(a) In which reaction is decomposition taking place? .... [1]

(b) In which reaction is combustion taking place? .... [1]

(c) In **two** reactions reduction is taking place.

Write down the letters for the **two** reactions and the substance which is being reduced in each case.

Reaction: .... Formula of substance reduced: .... [2]

Reaction: .... Formula of substance reduced: .... [2]

(d) The temperature in the blast furnace is kept up to a maximum of 1900 °C.

How is the temperature kept so high?

....

.... [2]

[Total 8 marks]

H38 Aluminium is extracted from purified aluminium oxide by electrolysis.
The diagram shows an electrolysis cell for extracting aluminium.

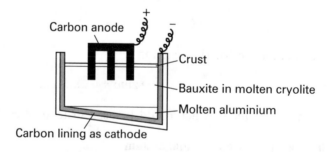

(a) Why is aluminium extracted from molten aluminium oxide dissolved in cryolite rather than molten aluminium oxide?

.... [1]

(b) Why do the carbon anodes have to be replaced frequently?

....

.... [2]

(c) (i) Finish the ionic equations to show the reaction taking place at the anode and cathode.

Cathode Al^{3+} + ... $\rightarrow$ ....

Anode   ....$O^{2-} \rightarrow$ .... + ... [4]

(ii) Explain why the reactions at the electrodes are oxidation and reduction.

Use your understanding of the definitions of oxidation and reduction in terms of electron transfer in your answer.

....

....

.... [3]

[Total 10 marks]

Formulae and equations

Word equations

A word equation is a useful summary of a chemical reaction. For example,
calcium carbonate + hydrochloric acid → calcium chloride + water + carbon dioxide

1 What name is given to the chemicals which react? .... [1]

2 What name is given to the chemicals which are formed? .... [1]

3 Here is a list of three chemicals: **barium nitrate**   **nitric acid**   **sulphuric acid**

Barium sulphate can be made by mixing two of these solutions together.
Fill the spaces with the names in the list to give a word equation.

.... + ... → barium sulphate + ... [2]

Symbol equations

You will find the following list of ions and their charges will help you to answer the questions:
Cl^- chloride OH^- hydroxide NO_3^- nitrate HCO_3^- hydrogencarbonate
Na^+ sodium H^+ hydrogen O^{2-} oxide SO_4^{2-} sulphate CO_3^{2-} carbonate
Ca^{2+} calcium Mg^{2+} magnesium Zn^{2+} zinc Pb^{2+} lead Al^{3+} aluminium

4 Below is a list of chemical formulae. Some are correct and some are incorrect.
For each formula, write the chemical name and put a tick if the formula is correct or a cross if it is incorrect. Finally write all the incorrect formulae correctly.

Formula	Name	Tick or cross
$NaCl_2$		
Na_2CO_3		
$NaSO_4$		
$NaNO_3$		
$CaOH_2$		
$CaHCO_3$		
$MgCl_2$		
$MgSO_4$		
$MgNO_3$		
Al_2SO_4		
HSO_4		
HCl		

Formula	Name	Tick or cross
ZnCl		
ZnOH		
$ZnCO_3$		
PbO		
$PbNO_3$		
PbOH		
$PbCl_2$		
$PbCO_3$		

[10]

Corrected formulae: ....

.... [11]

5 Iron is a transition metal. It can form iron(II), Fe^{2+}, and iron(III), Fe^{3+} ions.
Write down the formulae of:

(a) iron(II) sulphate ....   (b) iron(III) sulphate ....

(c) iron(II) hydroxide ....   (d) iron(III) oxide ....   [4]

6 What are the meaning of the following state symbols?

(s) ....   (l) ....

(g) ....   (aq) ....   [4]

7 Finish the following symbol equations. Make sure each equation is correctly balanced.

(a) $Mg(s) +$ $O_2(g) \rightarrow$ ....

(b) $Na(s) +$ $H_2O(l) \rightarrow$ $NaOH(aq) +$ ....

(c) $NaHCO_3(s) \rightarrow$ $Na_2CO_3(s) +$ $H_2O(l) +$ $CO_2(g)$

(d) $CH_4(g) +$ $O_2(g) \rightarrow$ $CO_2(g) +$ $H_2O(l)$

(e) $CaO(s) +$ $HNO_3(aq) \rightarrow$ $Ca(NO_3)_2(aq) +$ ....

(f) $MgCO_3(s) +$ $HCl(aq) \rightarrow$ $MgCl_2(aq) + H_2O(l) +$ ....

(g) $NH_3(g) +$ $O_2(g) \rightarrow$ $N_2(g) +$ $H_2O(g)$ [14]

8 Write balanced symbol equations using the information in these word equations.

(a) zinc + sulphuric acid $\rightarrow$ zinc sulphate + hydrogen

....

(b) aluminium oxide $\rightarrow$ aluminium + oxygen

....

(c) calcium hydroxide + carbon dioxide $\rightarrow$ calcium carbonate + water

.... [6]

Ionic equations are balanced in a similar way to other symbolic equations but also the sum of the charges on the left-hand side must be equal to the algebraic sum of the charges on the right-hand side. For example,

$$Fe^{2+} \rightarrow Fe^{3+} + e^-$$

Charges on LHS $= 2+$ Charges on RHS $= (3+) + (1-) = 2+$

9 Balance the following ionic equations:

(a) ___ $Cl^- \rightarrow$ ___ $Cl_2 +$ ___ e^- (b) ___ $Zn \rightarrow$ ___ $Zn^{2+} +$ ___ e^-

(c) ___ $O^{2-} +$ ___ $H^+ \rightarrow$ ___ H_2O [3]

10 When silver nitrate solution and sodium chloride solution are mixed, a white precipitate is formed.

(a) Write a balanced symbol equation for this reaction.

.... [2]

(b) Write an ionic equation for this reaction.

.... [2]

(c) What mass change, if any, would you expect. Explain your answer.

.... [2]

(d) Why would there be a mass change in the reaction below if it was carried out in an open beaker?

calcium carbonate + hydrochloric acid $\rightarrow$ calcium chloride + water + carbon dioxide

.... [1]

Practice Module Tests

Taking Module Tests

SEG Science Module Tests are 20 minutes long. You will have 18 questions on the test paper. Each question is worth one mark. There are two tiers or levels of papers. They are called Higher Tier and Foundation Tier.

Some questions appear on both papers. The Foundation Tier will have some easy questions and then the questions common to both papers. The Higher Tier paper has the common questions followed by some more difficult questions.

When you sit down to do the Module Test you will also have a computer answer sheet. You must write on this sheet, using an HB pencil, to record your answers. For each of the 18 questions you have to choose the best answer from the ones given. Your answer will be A, B, C, or D.

A sample question is

Which gas is taken out of the air by photosynthesis?
A. Carbon dioxide
B. Hydrogen
C. Nitrogen
D. Oxygen

The four possible answers are in alphabetical order. If the question involves answers that are numbers, they will be in numerical order.

The correct answer is **A**. This should be recorded on the answer sheet.

General advice

1. You must give only one answer to each question.
2. If you do not immediately know the answer to the question, leave it and come back to it.
3. If you decide to change an answer, use your rubber to rub out your original answer completely. If you do not rub it out properly, the computer will think you have given two answers.
4. If you need to do any working out, do it on the question paper and not on the answer sheet.
5. When you have finished all the questions, go back to any you left out.
6. Have a go at each question. There is no penalty if you get a question wrong. If you can definitely rule out some answers, you increase your chances if you make a guess.
7. Do not spend a lot of time going back and changing your answers. Usually your first answer is more likely to be correct.
8. Make sure that your name and examination number on your answer sheet are correct before you hand it in.

> **In this book, several questions are sometimes based on a single set of data. This has been done by the authors to fit in the maximum number of questions. It is not SEG policy to follow this practice. Only one SEG question is based on each set of data.**

Maintenance of life

1 Eagles eat rabbits. Which life process is this?
A. Growth
B. Nutrition
C. Respiration
D. Sensitivity

2 In which part of a cell do most chemical reactions take place?
A. Cell membrane
B. Chromosome
C. Cytoplasm
D. Nucleus

3 Which of the following is a type of tissue?
A. Kidney
B. Lung
C. Muscle
D. Stomach

4 Bacteria make us feel ill when they
A. divide inside the body
B. enter the body
C. multiply inside the body
D. produce toxins inside the body

Questions 5–7
The drawing shows human blood seen through a microscope.

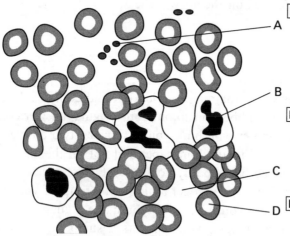

5 Which part of the blood, **A**, **B**, **C** or **D**, is a red cell?

6 Which part, **A**, **B**, **C** or **D**, is a platelet?

7 Which part, **A**, **B**, **C** or **D**, transports most oxygen around the body?

Questions 8–11
The drawing shows part of the digestion system.

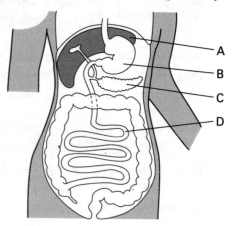

8 Which part, **A**, **B**, **C** or **D**, is the stomach?

9 Which part, **A**, **B**, **C** or **D**, is the liver?

10 In which part, **A**, **B**, **C** or **D**, does the digestion of protein begin?

11 In which part, **A**, **B**, **C** or **D**, is most soluble food absorbed into the blood?

H12 Which of the following organs produces the enzyme lipase?
A. Liver
B. Pancreas
C. Salivary gland
D. Stomach

H13 Bile is made in the
A. gall bladder
B. liver
C. small intestine
D. stomach

H14 Bile breaks down
A. fat drops into amino acids
B. fat drops into fat droplets
C. fat drops into fatty acids
D. fat drops into glucose

H15 The liver stores excess sugar as
A. cellulose
B. glucose
C. glycogen
D. starch

Questions 16–18

The drawing shows the front of the eye.

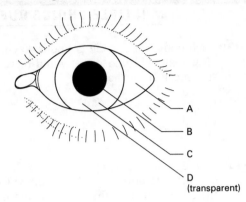

16 Which part, **A**, **B**, **C** or **D**, is the cornea?

17 Which part, **A**, **B**, **C** or **D**, is the iris?

18 Which part, **A**, **B**, **C** or **D**, is the pupil?

19 Which part of the eye is transparent?
A. Ciliary muscle
B. Cornea
C. Iris
D. Retina

20 Which part of the eye controls the amount of light that enters?
A. Ciliary muscle
B. Cornea
C. Iris
D. Retina

21 Which part of the eye contains receptors sensitive to light?
A. Ciliary muscle
B. Cornea
C. Iris
D. Retina

Questions 22–24

The drawing shows some of the organs concerned with maintaining a constant interval environment.

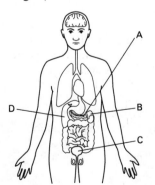

22 Which organ, **A**, **B**, **C** or **D**, produces urine?

23 Which organ, **A**, **B**, **C** or **D**, produces insulin?

H24 Which organ, **A**, **B**, **C** or **D**, stores glycogen?

25 If we smell food our salivary glands automatically produce saliva. In this reflex action the salivary gland is the
A. effector
B. neurone
C. receptor
D. stimulus

26 Which is a hormone concerned with controlling blood sugar levels?
A. ADH
B. Glucose
C. Glycogen
D. Insulin

27 At the end of a race a runner's face was very red. This was due to
A. constriction of the blood vessels supplying the skin capillaries
B. decreased sweating
C. dilation of the blood vessels supplying the skin capillaries
D. increased sweating

28 We produce less urine on a hot day because
A. the bladder can hold more urine
B. the kidneys do not work as fast
C. we drink less
D. we sweat more

1	A	B	C	D	11	A	B	C	D	21	A	B	C	D
2	A	B	C	D	12	A	B	C	D	22	A	B	C	D
3	A	B	C	D	13	A	B	C	D	23	A	B	C	D
4	A	B	C	D	14	A	B	C	D	24	A	B	C	D
5	A	B	C	D	15	A	B	C	D	25	A	B	C	D
6	A	B	C	D	16	A	B	C	D	26	A	B	C	D
7	A	B	C	D	17	A	B	C	D	27	A	B	C	D
8	A	B	C	D	18	A	B	C	D	28	A	B	C	D
9	A	B	C	D	19	A	B	C	D					
10	A	B	C	D	20	A	B	C	D					

1 A bacterium reproduces by dividing into two. This is an example of
A. asexual reproduction
B. fertilisation
C. mutation
D. sexual reproduction

2 Identical twins are formed when
A. a fertilised egg divides into two cells which separate.
B. an egg cell divides into two before being fertilised.
C. one egg is fertilised by two sperms.
D. two eggs are fertilised at the same time.

3 The stomach cells of a girl have
A. one X chromosome and one Y chromosome
B. one X chromosome only
C. two X chromosomes
D. two Y chromosomes

4 For a boy to be produced the fertilising sperm will have
A. one X chromosome
B. one Y chromosome
C. two X chromosomes
D. two Y chromosomes

H5 During which of the following processes will the number of chromosomes be halved?
A. a bacterium dividing into two
B. cloning plant cells
C. cuttings growing into a plant
D. producing pollen grains in an anther

H6 Sexual reproduction leads to more variation than asexual reproduction because
A. genes pair up in new ways during fertilisation
B. mutations only occur during sexual reproduction
C. natural selection takes place
D. only males have Y chromosomes

7 Gardeners often grow chrysanthemums from cuttings rather than seeds. The advantage of this is that
A. all the flowers will be the same colour
B. new varieties of chrysanthemums are produced
C. the new plants will flower quicker
D. the new plants grow bigger

8 If there is a leak of radioactive materials from a nuclear power station
A. mutation rates will increase
B. rivers will have less oxygen
C. the atmosphere will get warmer
D. the rain will become more acidic

H9 Each body cell of a cat contains 40 chromosomes. How many chromosomes would there be in a sperm cell from the cat?
A. 20
B. 23
C. 40
D. 46

H10 A disease is caused by a recessive allele c. A man and his wife both have the genotype Cc. What is the chance of their first child having the disease?
A. Nil
B. 1 in 2
C. 1 in 3
D. 1 in 4

H11 In one plant species red-flower colour is dominant and white-flower colour is recessive. When a red-flowered plant was crossed with another red-flowered plant 25 per cent of the offspring had white flowers. Which one of the following best describes the pollen of the red-flowered parent plant?
A. 25% of the pollen grains had the allele for red-flower, 75% had the allele for white-flower
B. 50% of the pollen grains had the allele for red-flower, 50% had the allele for white-flower
C. 75% of the pollen grains had the allele for red-flower, 25% had the allele for white-flower
D. All the pollen grains had the allele for red-flower

H12 In cats the dominant allele of a gene, M, results in tail-less animals. The recessive allele, m, produces cats with a tail. Cats which are homozygous for the dominant allele are never produced because they die in the uterus at a very early stage. If two tail-less cats are mated, the proportion of cats with tails in the offspring will be
A. 1/4
B. 1/3
C. 2/3
D. 3/4

H13 All the plants in a population have red flowers. They are found to be homozygous for the allele which produces red flowers. Flowers of a different colour could arise in this population only as a result of
A. artificial selection
B. asexual reproduction
C. mutation
D. natural selection

H14 Some types of disease-causing bacteria have become resistant to some antibiotics. This is the result of
A. the antibiotics causing mutations in the bacteria
B. the antibiotics not being powerful enough
C. the bacteria becoming used to antibiotics
D. the survival of varieties of bacteria which are not affected by the antibiotic

15 Which of the following is most likely to damage the liver?
A. Drinking alcohol
B. Lack of exercise
C. Lack of fibre in the diet
D. Smoking tobacco

16 Which of the following gases dissolves to form acid rain?
A. Carbon monoxide
B. Nitrogen
C. Oxygen
D. Sulphur dioxide

17 Penguins use their wings for swimming instead of flying. This is an example of
A. adaptation
B. competition
C. continuous variation
D. mutation

18 Which of the following is due to there being two alleles of a gene for a character?
A. Continuous variation
B. Discontinuous variation
C. Mutation
D. Natural selection

19 Which of the following is shown by the human character height?
A. Continuous variation
B. Discontinuous variation
C. Mutation
D. Natural selection

H20 Which of the following is due to one allele of a gene producing features which favour survival more than the other allele of the gene?
A. Continuous variation
B. Discontinuous variation
C. Mutation
D. Natural selection

H21 The table shows the useful characteristics of three varieties of cattle.

Variety	Average milk yield	Beef production	Survival in hot weather
1	low	good	good
2	low	good	poor
3	high	poor	poor

A breeder was asked to provide cattle for a dairy farm in a hot country. Which two varieties should she breed from?
A. 1 and 2
B. 1 and 3
C. 2 and 3
D. 3 and 3

22 The graph shows a change in the population of animals.

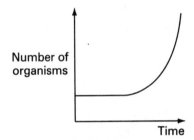

This change was probably due to
A. colder weather
B. increased competition for territory
C. reduced food supplies
D. reduced predation

H23 In industrial areas soot darkens the trees. Here, more dark moths than light moths are found on tree trunks. In clean countryside more light moths than dark moths are found. The best explanation for this is that
A. birds can see light moths more easily on sooty trees
B. dark moths fly to polluted areas for camouflage
C. moths can change colour to match their surroundings
D. moths darken as they get older

H24 A plant breeder may produce new varieties of potato plants by several different methods. Which of the following methods would give offspring with the widest range of genotypes?
A. Cloning the parent plant cells
B. Cross-pollination of the parent plant
C. Planting of cuttings from the parent plant
D. Self-pollination of the parent plant

H25 When a lake is polluted by excess fertilisers, the oxygen supply in the lake decreases. This is mainly the result of
A. an increase in the population of fish
B. an increase in the population of microbes
C. oxygen combining with the fertilisers
D. photosynthesis by the green plants

1	A	B	C	D	10	A	B	C	D	19	A	B	C	D
2	A	B	C	D	11	A	B	C	D	20	A	B	C	D
3	A	B	C	D	12	A	B	C	D	21	A	B	C	D
4	A	B	C	D	13	A	B	C	D	22	A	B	C	D
5	A	B	C	D	14	A	B	C	D	23	A	B	C	D
6	A	B	C	D	15	A	B	C	D	24	A	B	C	D
7	A	B	C	D	16	A	B	C	D	25	A	B	C	D
8	A	B	C	D	17	A	B	C	D					
9	A	B	C	D	18	A	B	C	D					

Structure and changes

Questions 1–4

The elements in Group 7 are shown in the table.

Element	Atomic number
bromine	35
chlorine	17
fluorine	9
iodine	53

1 Which element has the largest atomic radius?
A. Bromine
B. Chlorine
C. Fluorine
D. Iodine

2 Which element is most reactive?
A. Bromine
B. Chlorine
C. Fluorine
D. Iodine

3 Which element has the highest melting and boiling point?
A. Bromine
B. Chlorine
C. Fluorine
D. Iodine

4 Which element is liquid at room temperature and atmospheric pressure?
A. Bromine
B. Chlorine
C. Fluorine
D. Iodine

H5 Which of the following are **not** produced on electrolysis of sodium chloride solution (brine)?
A. Chlorine
B. Hydrogen
C. Sodium
D. Sodium hydroxide

6 Which noble gas is used to fill balloons?
A. Argon
B. Helium
C. Krypton
D. Neon

7 The gas produced when sodium is added to cold water is
A. carbon dioxide
B. hydrogen
C. oxygen
D. steam

8 The table gives information about alkali metal halides.
Which row of the table is correct?

	Type of bonding	Solubility in water
A	ionic	poor
B	ionic	good
C	covalent	poor
D	covalent	good

H9 The table gives information about alkali metal hydroxides.
Which row of the table is correct?

	Type of bonding	pH of solution	Ion present
A	ionic	1	H^+
B	ionic	13	OH^-
C	covalent	1	H^+
D	covalent	13	OH^-

10 The elements in the modern Periodic Table are arranged in order of
A. date of discovery
B. increasing atomic mass
C. increasing atomic number
D. increasing atomic radius

11 Which is the positive test for hydrogen?
A. A glowing splint relights
B. A lighted splint goes out
C. A lighted splint produces a squeaky pop
D. Limewater turns milky

12 Which of the following does **not** produce hydrogen gas when heated with hydrochloric acid?
A. Copper
B. Iron
C. Magnesium
D. Zinc

13 Which of the following metals will react when a mixture of the metal powder and zinc oxide are heated together?
A. Copper
B. Gold
C. Iron
D. Magnesium

14 Aluminium does not corrode as quickly as iron because
 A. an insoluble oxide forms on the surface preventing reaction
 B. it is a less reactive metal
 C. it is below iron in the reactivity series
 D. it is protected by sacrificial protection

15 Iron filings, zinc powder and magnesium powder are added to separate samples of blue copper(II) sulphate solution. How many of the solutions lose their blue colour?
 A. None
 B. One
 C. Two
 D. Three

16 Lime is added to soil
 A. as a fertiliser
 B. to kill pests
 C. to neutralise acid in the soil
 D. to neutralise alkali in the soil

Questions 17–20
The table shows the pH values of four solutions **A**, **B**, **C** and **D**.

	A	B	C	D
pH	4	11	7	1

17 Which solution is exactly neutral?

18 Which solution is a strong alkali?

19 Which solution is a weak acid?

20 Which solution would neutralise a solution with a pH of 2?

21 When dilute hydrochloric acid and sodium thiosulphate solution are mixed the solution
 A. gets hot
 B. gives off bubbles of colourless gas
 C. goes blue
 D. goes cloudy

22 Which of the following will speed up the reaction between calcium carbonate and hydrochloric acid?
 A. Adding water to the hydrochloric acid
 B. Breaking the lumps of calcium carbonate into small lumps
 C. Reducing the temperature
 D. Using half the mass of calcium carbonate

Questions 23–27
The graph shows the results of an experiment involving the decomposition of 50 cm^3 of hydrogen peroxide solution. Manganese oxide is added.

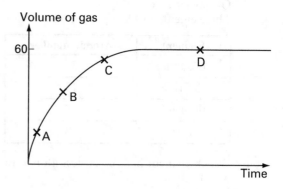

23 The manganese oxide acts as
 A. a catalyst
 B. an oxidising agent
 C. a reactant
 D. a reducing agent

24 At which point **A**, **B**, **C** or **D** is the reaction fastest?

25 At which point **A**, **B**, **C** or **D** is the reaction finished?

26 If the experiment was repeated using 25 cm^3 of the same hydrogen peroxide solution, what would be the final volume of oxygen collected?
 A. 30 cm^3
 B. 45 cm^3
 C. 60 cm^3
 D. 120 cm^3

H27 A reaction takes place between two gases.
Which of the following will reduce the number of effective collisions?
 A. Adding some noble gas to the mixture of gases while keeping the pressure unchanged
 B. Increasing the concentration of the gases
 C. Increasing the pressure
 D. Raising the temperature of the reaction

28 Which is the best test for carbon dioxide?
 A. Puts out a lighted splint
 B. Relights a glowing splint
 C. Turns limewater milky
 D. Turns litmus paper red

29 The steps in making plastics from crude oil are
1. cracking
2. fractional distillation
3. polymerisation

The correct order of these steps is
A. 1, 2, 3
B. 2, 1, 3
C. 2, 3, 1
D. 3, 2, 1

H30 Which of the following is the correct displayed structural formula for ethene?

A   **B**

C   **D**

H31 Alkanes and alkenes can be distinguished with
A. a lighted splint
B. bromine water
C. limewater
D. litmus paper

32 Alkanes have a general formula C_nH_{2n+2}.
Which of the following is an alkane?
A. C_2H_2
B. C_3H_6
C. C_4H_{10}
D. C_6H_6

33 Which of the following is not a hydrocarbon?
A. Methane, CH_4
B. Ethanol, C_2H_5OH
C. Ethene, C_2H_4
D. Ethane, C_2H_6

H34 An unsaturated hydrocarbon
A. contains a double carbon–carbon bond
B. contains only single carbon–carbon bonds
C. does not burn in air
D. will dissolve in water

35 Which of the following is a poisonous gas sometimes produced when methane burns in a faulty gas fire?
A. carbon dioxide
B. carbon monoxide
C. hydrogen
D. sulphur dioxide

1	A	B	C	D	13	A	B	C	D	25	A	B	C	D
2	A	B	C	D	14	A	B	C	D	26	A	B	C	D
3	A	B	C	D	15	A	B	C	D	27	A	B	C	D
4	A	B	C	D	16	A	B	C	D	28	A	B	C	D
5	A	B	C	D	17	A	B	C	D	29	A	B	C	D
6	A	B	C	D	18	A	B	C	D	30	A	B	C	D
7	A	B	C	D	19	A	B	C	D	31	A	B	C	D
8	A	B	C	D	20	A	B	C	D	32	A	B	C	D
9	A	B	C	D	21	A	B	C	D	33	A	B	C	D
10	A	B	C	D	22	A	B	C	D	34	A	B	C	D
11	A	B	C	D	23	A	B	C	D	35	A	B	C	D
12	A	B	C	D	24	A	B	C	D					

1 Which device is used to measure electric current?
A. Ammeter
B. Joulemeter
C. Voltmeter
D. Wattmeter

Questions 2 and 3
The graph shows how the distance travelled by a moving object changes with time.

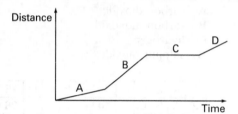

2 Which part of the graph shows that the object was stationary?

3 Which part of the graph represents the fastest speed?

4 The diagram shows reflection of light at a mirror.

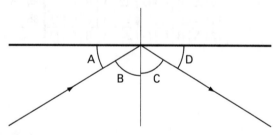

Which pair of marked angles are equal in size?
A. A and B
B. A and C
C. B and C
D. B and D

5 Which type of electromagnetic radiation is used for the remote control of televisions?
A. Infrared
B. Microwaves
C. Radio waves
D. Ultraviolet

6 Resistance is measured in
A. amps
B. ohms
C. volts
D. watts

Questions 7–10
The diagram shows a correctly wired three-pin plug.

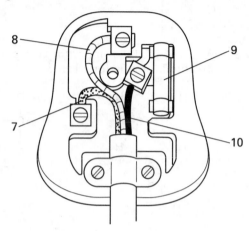

Choose the correct name from this list for each of the labelled parts.
A. Earth wire
B. Fuse
C. Live wire
D. Neutral wire

11 Which statement about alternating current and direct current is correct?
A. Direct current can only pass from negative to positive but alternating current can pass in either direction
B. Direct current only passes in one direction but alternating current changes direction
C. Direct current changes direction but alternating current only passes in one direction
D. Direct current comes from the mains supply but alternating current comes from batteries

12 Which statement about a fuse is correct?
A. The fuse limits the amount of current that passes in a circuit
B. The fuse melts if the current becomes too great
C. The fuse provides a low-resistance path to earth
D. The fuse is always placed in the neutral connection

13 A sprinter completes an 80 m race in 10 s. The sprinter's average speed is:
A. 8 m/s
B. 8 m/s^2
C. 800 m/s
D. 800 m/s^2

14 Which statement about mass and weight is correct?
A. Mass is a force and weight is not a force
B. Weight is measured in kilograms; mass is measured in newtons
C. Weight is measured in newtons; mass is measured in kilograms
D. Weight is a force and mass is a force

15 A force is used in an attempt to open a door.

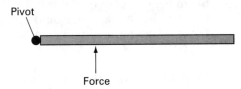

The door does not open.
Which of the following could cause the door to open?
A. Changing the angle at which the force is applied
B. Moving the force further away from the pivot
C. Moving the force nearer to the pivot
D. Using a smaller force

16 An object falling through the Earth's atmosphere moves at terminal velocity if
A. air resistance is negligible
B. air resistance is smaller than the weight
C. air resistance is greater than the weight
D. air resistance is equal to the weight

17 A spring follows Hooke's law if
A. the extension of the spring is proportional to the force
B. the length of the spring is proportional to the force
C. when the force on the spring increases, the extension increases
D. when the force on the spring increases, the extension decreases

Questions 18–19
The diagram shows part of a surface water wave.

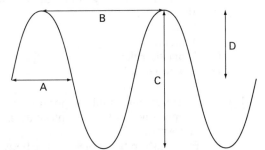

18 Which distance on the diagram shows the amplitude?

19 Which distance on the diagram shows the wavelength?

20 When an X-ray photograph is taken, the X-rays
A. are reflected by bone but not by skin
B. are reflected by skin but not by bone
C. pass through bone but not through flesh
D. pass through flesh but not through bone

21 Which diagram shows correctly the path of light through a window?

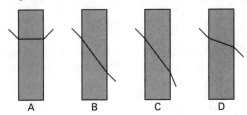

22 Which list shows radio waves, light, infrared and ultraviolet radiation in order of increasing wavelength?
A. Infrared, ultraviolet, radio, light
B. Light, infrared, ultraviolet, radio
C. Radio, ultraviolet, light, infrared
D. Ultraviolet, light, infrared, radio

23 When a magnet is moved into a coil of wire a voltage is created in the coil.
Which action would **not** cause an increase in the size of the voltage?
A. Moving the magnet at the same speed out of the coil
B. Moving the magnet faster
C. Using a coil of the same size with more turns of wire
D. Using a magnet with a stronger magnetic field

Questions 24 and 25
The resistance of a kettle element is 24 Ω. It is connected to 240 V mains.

24 The current rating of the fuse fitted to the plug is
A. 1 A
B. 3 A
C. 5 A
D. 13 A

25 The power of the element is
A. 24 W
B. 240 W
C. 2400 W
D. 24 000 W

1	A	B	C	D	10	A	B	C	D	19	A	B	C	D
2	A	B	C	D	11	A	B	C	D	20	A	B	C	D
3	A	B	C	D	12	A	B	C	D	21	A	B	C	D
4	A	B	C	D	13	A	B	C	D	22	A	B	C	D
5	A	B	C	D	14	A	B	C	D	23	A	B	C	D
6	A	B	C	D	15	A	B	C	D	24	A	B	C	D
7	A	B	C	D	16	A	B	C	D	25	A	B	C	D
8	A	B	C	D	17	A	B	C	D					
9	A	B	C	D	18	A	B	C	D					

Energy sources

1 Which energy source is non-renewable?
A. Coal
B. Waves
C. Wind
D. Wood

2 A solar panel can be used to heat water.
The energy that heats the water comes from
A. electricity
B. gas
C. the Moon
D. the Sun

3 Heat (thermal energy) flows through a solid metal mainly by
A. conduction
B. convection
C. evaporation
D. radiation

4 Radioactivity can be detected by
A. an ammeter
B. a Geiger–Müller tube
C. a voltmeter
D. an X-ray tube

5 Which object is a star?
A. Galaxy
B. Moon
C. Planet
D. Sun

6 Which type of electromagnetic radiation is emitted by food taken from a warm oven?
A. Gamma rays
B. Infrared radiation
C. Light
D. Ultraviolet radiation

7 One form of loft insulation consists of fibres of glass. Which statement best describes why glass fibre is an effective insulator?
A. Air is trapped between the fibres
B. Heat is trapped in the fibres
C. The fibres only allow cold air to pass through them
D. The fibres reflect thermal energy

8 Which statement about cavity wall insulation is correct?
A. It keeps warm air in and cold air out
B. It keeps warm air out and cold air in
C. It prevents energy transfer by radiation
D. It prevents convection currents between the outer wall and the inner wall

9 After running a race, marathon runners are sometimes given a cape made out of aluminium foil. The purpose of the cape is
A. to allow heat to escape by conduction
B. to help them to cool down
C. to reflect hot air back onto their bodies
D. to reflect infrared radiation back onto their bodies

10 Domestic smoke alarms contain radioactive sources that emit alpha particles.
They do not present a hazard to humans because
A. alpha particles do not affect human tissue
B. alpha particles are electromagnetic waves
C. alpha particles cannot penetrate the casing of the smoke alarm
D. alpha particles are not charged

11 The diagram shows how a radioactive source is used in the production of aluminium foil.

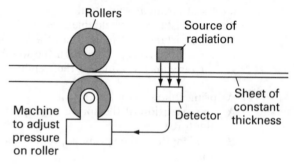

The radioactive source used should emit
A. alpha particles only
B. alpha particles and gamma radiation
C. beta particles and gamma radiation
D. beta particles only

12 Which of the following is most likely to be very slightly radioactive?
A. A glass beaker
B. A metal spoon
C. A plastic ruler
D. A wooden table

13 Which statement about evaporation is true?
A. Evaporation only takes place at one temperature
B. Evaporation takes place in the body of a liquid
C. Evaporation takes place from the surface of a liquid
D. Liquids evaporate more slowly at higher temperatures

14 The Solar System is part of which galaxy?
A. Andromeda
B. Magellanic Clouds
C. Milky Way
D. Sombrero

15 The diagram represents the innermost planets.

Mars

Sun ● Mercury

Earth

Which inner planet is missing from the diagram?
A. Jupiter
B. Neptune
C. Pluto
D. Venus

16 The Moon orbits the Earth because of the
A. attractive gravitational force between them
B. attractive magnetic force between them
C. repulsive gravitational force between them
D. repulsive magnetic force between them

17 Seasons are caused by
A. the Earth's axis being tilted
B. the Earth's rotation around the Sun
C. the Earth spinning on its own axis
D. the Moon's rotation around the Earth

18 Which planet takes the longest time to orbit the Sun?
A. Neptune
B. Pluto
C. Saturn
D. Uranus

H19 The main reaction taking place in the Sun is
A. the fission of helium nuclei into hydrogen nuclei
B. the fission of hydrogen nuclei into helium nuclei
C. the fusion of helium nuclei into hydrogen nuclei
D. the fusion of hydrogen nuclei into helium nuclei

H20 Radon is a radioactive gas with a half-life of 55 s. The activity of a sample of radon is measured to be 80 counts/s.
After 110 s the activity is likely to be
A. 0 count/s
B. 10 count/s
C. 20 count/s
D. 40 count/s

21 A patient is to be injected with a radioactive isotope to show the blood flow through the liver. Which isotope is most suitable?
A. An alpha emitter with a half-life of 6 hours
B. A beta emitter with a half-life of 6 days
C. A gamma emitter with a half-life of 6 hours
D. A gamma emitter with a half-life of 6 days

22 Which of the following uses of a satellite is most likely to use a geosynchronous satellite?
A. An orbiting space station
B. Monitoring the weather
C. Observing the Solar System
D. Television communications

23 Which statement about a comet is true?
A. It maintains a constant speed in its orbit
B. It orbits in the same plane as the planets
C. Its orbit is a circle
D. Its orbit is an ellipse

24 Which statement about an exothermic reaction is correct?
A. Energy is released when bonds are broken
B. More energy is produced from bond formation than is required for bond breaking
C. No bonds are broken and no new bonds are formed
D. The products contain more energy than the reactants

25 Here are two chemical reactions:

1. Copper(II) sulphate crystals are heated.
2. Water is added to anhydrous copper(II) sulphate.

Which of these reactions are exothermic?
A. 1 only
B. 1 and 2
C. 2 only
D. Neither

1	A	B	C	D	10	A	B	C	D	19	A	B	C	D
2	A	B	C	D	11	A	B	C	D	20	A	B	C	D
3	A	B	C	D	12	A	B	C	D	21	A	B	C	D
4	A	B	C	D	13	A	B	C	D	22	A	B	C	D
5	A	B	C	D	14	A	B	C	D	23	A	B	C	D
6	A	B	C	D	15	A	B	C	D	24	A	B	C	D
7	A	B	C	D	16	A	B	C	D	25	A	B	C	D
8	A	B	C	D	17	A	B	C	D					
9	A	B	C	D	18	A	B	C	D					

Vital exchanges

1 The chloroplasts in a plant cell
  A. absorb light energy
  B. contain chromosomes
  C. control the entry of materials into the cell
  D. strengthen the cell

2 The cell wall in a plant cell
  A. absorbs light energy
  B. stores starch
  C. controls the entry of materials into the cell
  D. strengthens the cell

3 The main function of the vacuole in a plant cell is
  A. diffusion
  B. photosynthesis
  C. respiration
  D. support

4 Which of the following is **not** found in animal cells?
  A. Cell membrane
  B. Cell wall
  C. Cytoplasm
  D. Nucleus

5 Air passes from the throat towards the lungs through the
  A. diaphragm
  B. intestine
  C. oesophagus
  D. trachea

6 Oxygen diffuses into the blood capillaries through the walls of the
  A. alveoli
  B. diaphragm
  C. intestine
  D. trachea

7 All arteries
  A. carry blood away from the heart
  B. carry oxygenated blood
  C. have valves
  D. have walls one cell thick

8 Which of the following is a product of photosynthesis?
  A. Carbon dioxide
  B. Chlorophyll
  C. Glucose
  D. Water

Questions 9 and 10
Metal foil was wrapped round one leaf of a potted plant as shown in the drawing.

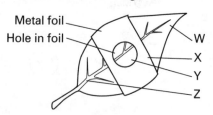

The plant was left in bright light for 4 hours. The leaf was then removed and tested for glucose.

9 Which parts of the leaf **W**, **X**, **Y** and **Z**, would give a positive result in the glucose test?
  A. X, Y and Z
  B. W, X and Y
  C. W, X, Y and Z
  D. W, Y and Z

10 The reason for this was that some parts of the leaf did not receive
  A. carbon dioxide
  B. light
  C. oxygen
  D. water

11 A student set up the apparatus shown in the diagram.

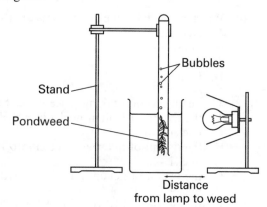

The lamp was set up at different distances from the pondweed. At each distance the student counted the number of bubbles given off in one minute. The experiment was repeated three times and the mean number of bubbles given off per minute was calculated.

The results are shown in the table.

Distance from lamp to pondweed in cm	Mean number of bubbles given off per minute
5	61
10	20
15	9
20	4
25	3
30	3

Which of the following is the best explanation of these results?
A. At high light intensities some other factor is limiting the rate of photosynthesis
B. At low light intensities light is the limiting factor for photosynthesis
C. The higher the light intensity the higher the rate of photosynthesis
D. The rate of photosynthesis depends entirely on light intensity

12 Which of the following would increase the rate of transpiration?
A. Darkness
B. High humidity
C. High temperature
D. Low wind speed

13 Which of the following might cause plants to wilt?
A. A cold day
B. A still day
C. A well-watered soil
D. The soil being flooded by sea water

H14 The main method used by root hair cells to absorb water from the soil is
A. active uptake
B. diffusion
C. osmosis
D. transpiration

H15 The main method used by root hair cells to absorb mineral ions from the soil is
A. active transport
B. diffusion
C. osmosis
D. transpiration

16 Carbon dioxide may be removed from the atmosphere by
A. animals
B. burning oil
C. decaying microbes
D. green plants

H17 A freshly cut piece of potato was placed in a concentrated solution of common salt. After 4 hours it was soft and flabby. This was because
A. salt had passed from the potato cells into the salt solution
B. salt had passed from the salt solution into the potato
C. water had passed from the potato cells into the salt solution
D. water had passed from the salt solution into the potato

18 Which of the following is a waste product of aerobic respiration?
A. carbon dioxide
B. energy
C. glucose
D. oxygen

19 A student decided to investigate how long she could do continuous step-ups. She found that after six minutes she could do no more.
This was because
A. her body had run out of glucose
B. lactic acid had built up in her muscles
C. she was breathing too fast
D. there was too much carbon dioxide in her blood

20 All the energy which flows through a food chain comes originally from
A. photosynthesis
B. respiration
C. sunlight
D. the burning of coal

21 In the food chain
grass → rabbits → foxes
the foxes are feeding as
A. carnivores
B. decomposers
C. herbivores
D. producers

H22 Nitrogen-fixing bacteria
A. convert ammonium compounds into nitrate
B. make nitrogen compounds from atmospheric nitrogen
C. release nitrogen gas into the atmosphere
D. use nitrogen compounds as a source of energy

23 An oak wood has 150 trees. Many thousands of insects feed on their leaves. About 50 birds live in the wood and feed on the insects. Which pyramid, **A**, **B**, **C** or **D**, is the pyramid of biomass for these organisms?

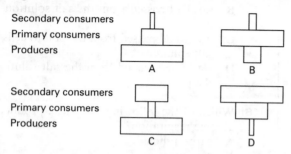

Secondary consumers
Primary consumers
Producers

A   B

Secondary consumers
Primary consumers
Producers

C   D

1	A	B	C	D		9	A	B	C	D		17	A	B	C	D
2	A	B	C	D		10	A	B	C	D		18	A	B	C	D
3	A	B	C	D		11	A	B	C	D		19	A	B	C	D
4	A	B	C	D		12	A	B	C	D		20	A	B	C	D
5	A	B	C	D		13	A	B	C	D		21	A	B	C	D
6	A	B	C	D		14	A	B	C	D		22	A	B	C	D
7	A	B	C	D		15	A	B	C	D		23	A	B	C	D
8	A	B	C	D		16	A	B	C	D						

Questions 1–2
A fluorine atom contains 9 protons, 9 electrons and 10 neutrons.

1 Which row in the table is correct?

	Atomic number	Mass number
A	9	19
B	9	18
C	10	19
D	10	18

2 The electron arrangement in a fluorine atom is
A. 2, 2, 5
B. 2, 7
C. 2, 8
D. 7, 2

3 An ionic bond involves
A. one metal atom and one non-metal atom
B. sharing a pair of electrons
C. two metal atoms
D. two non-metal atoms

H4 Which of the following does **not** contain one or more double covalent bonds?
A. Carbon dioxide
B. Ethene
C. Oxygen
D. Water

H5 The electron arrangements in aluminium and fluorine atoms are 2, 8, 3 and 2, 7 respectively. Which ions are present in aluminium fluoride?
A. Al^+ and F^-
B. Al^{2+} and F^-
C. Al^{3+} and F^-
D. Al^{3+} and F^{2-}

H6 A manganese atom contains 25 protons, 25 electrons and 30 neutrons. It can be written as
A. $^{55}_{25}Mn$ B. $^{25}_{55}Mn$ C. $^{55}_{30}Mn$ D. $^{50}_{25}Mn$

H7 There are two isotopes of chlorine: chlorine-35 and chlorine-37.
Which of the statements is true?
Both isotopes contain the same number of
A. neutrons
B. neutrons and electrons
C. protons and electrons
D. protons and neutrons

8 A magnesium ion, Mg^{2+}, contains
A. two fewer electrons than a magnesium atom
B. two fewer protons than a magnesium atom
C. two more electrons than a magnesium atom
D. two more protons than a magnesium atom

9 An element has atoms with an electronic structure of 2, 8, 6.
Which of the following is **not** true?
A. The atoms of the element each gain two electrons when forming ions
B. The element is a metal
C. The element is in Group 6 of the Periodic Table
D. The element is in Period 3 of the Periodic Table

10 Sulphur has an atomic number of 16 and a mass number of 32.
The electronic structure of a sulphur atom is
A. 2, 6, 8
B. 2, 8, 6
C. 2, 8, 18, 4
D. 8, 8

Questions 11–14
The table gives numbers of protons, neutrons and electrons in four atoms or ions, labelled **A**, **B**, **C** and **D**.

	Protons	Neutrons	Electrons
A	7	8	7
B	8	8	8
C	8	10	8
D	9	10	10

11 Which one is an ion?

12 Which one has a mass number of 18?

13 Which one has an atomic number of 7?

H14 Which one is an isotope of $^{14}_{7}N$?
A. P   B. Q   C. R   D. S

15 What is the relative formula mass of potassium hydrogencarbonate ($KHCO_3$)?
(Relative atomic masses: H = 1, C = 12, O = 16, K = 39.)
A. 68   B. 88   C. 100   D. 112

16 What mass of calcium hydroxide, $Ca(OH)_2$, contains 40 g of calcium?
(Relative atomic masses: $Ca = 40$.)
A. 57 g
B. 58 g
C. 74 g
D. 114 g

17 Which element is present in the largest amount, by mass, in potassium hydrogencarbonate ($KHCO_3$)?
(Relative atomic masses: $H = 1$, $C = 12$, $O = 16$, $K = 39$.)
A. Carbon
B. Hydrogen
C. Oxygen
D. Potassium

18 When 2.4 g of magnesium burns in oxygen, 4.0 g of magnesium oxide is formed. What mass of oxygen combines with 2.4 g of magnesium.
A. 1.6 g
B. 2.4 g
C. 4.0 g
D. 5.6 g

19 What percentage of sulphur is present in sulphur dioxide (SO_2)?
(Relative atomic masses: $O = 16$, $S = 32$.)
A. 32%   B. 50%   C. 64%   D. 80%

20 Which of the following hydrocarbons contain the greatest percentage of carbon?
(Relative atomic masses: $H = 1$, $C = 12$.)
A. CH_4 B. C_2H_6 C. C_2H_4 D. C_2H_2

Questions 21–23
The equation for the combustion of methane in excess oxygen is

$$CH_4(g) + 2O_2(g) \rightarrow CO_2(g) + 2H_2O(l)$$

One mole of any gas occupies 24 000 cm^3 at room temperature and atmospheric pressure.

All volumes are measured at room temperature and atmospheric pressure.

H21 What is the volume, in cm^3, of 2 moles of oxygen at room temperature and pressure?
A. 2   B. 12 000   C. 24 000   D. 48 000

H22 What is the minimum volume of oxygen, in cm^3, needed to burn 10 cm^3 of methane?
A. 10   B. 20   C. 30   D. 40

H23 What is the volume of the products, in cm^3, when 10 cm^3 of methane and 30 cm^3 of oxygen are burned?
A. 10   B. 20   C. 30   D. 40

H24 Potassium hydroxide has a formula mass of 40 g. How many moles of potassium hydroxide are present in 20 g of potassium hydroxide?
A. 0.5   B. 2   C. 40   D. 60

25 A sample of 30 g of a nitrogen oxide contains 14 g of nitrogen.
What is the formula of the nitrogen oxide?
(Relative atomic masses: $N = 14$, $O = 16$.)
A. NO B. NO_2 C. N_2O D. N_2O_4

26 Which of the following is likely to be a transition metal?

	Density of metal	Colour of compounds
A	high	coloured
B	high	not coloured
C	low	coloured
D	low	not coloured

H27 The catalyst in the Haber process is
A. aluminium
B. aluminium oxide
C. iron
D. platinum

28 In the Haber process, nitrogen and hydrogen are mixed.
Which row of the table gives the correct proportions of the two gases, by volume?

	Nitrogen	Hydrogen
A	1	3
B	3	1
C	1	1
D	1	2

29 The raw materials in the Haber process are
A. limestone, air and water
B. air, water and natural gas
C. air, natural gas and limestone
D. water, natural gas and limestone

H30 Which row of the table gives the best conditions of temperature and pressure for the Haber process?

	Temperature	Pressure
A	high	high
B	high	low
C	low	high
D	low	low

31 In the Haber process about 10% of the nitrogen/hydrogen mixture is converted to ammonia. The ammonia is removed by
A. compressing
B. distilling
C. filtering
D. liquefying

32 In the Haber process the unreacted nitrogen and hydrogen are
A. burnt
B. recycled
C. released into the atmosphere
D. separated

33 Which of the following is **not** a transition metal?
A. Copper
B. Magnesium
C. Manganese
D. Zinc

34 Which of the following properties is most important for a metal used for hot water pipes?
A. Does not react with hot water
B. High density
C. High melting point
D. Shiny

35 Which of the following is a reversible reaction?
A. Reaction of ammonia and nitric acid
B. Reaction of ammonia and oxygen
C. Reaction of ammonia and sulphuric acid
D. Reaction of nitrogen and hydrogen

1	A	B	C	D	13	A	B	C	D	25	A	B	C	D
2	A	B	C	D	14	A	B	C	D	26	A	B	C	D
3	A	B	C	D	15	A	B	C	D	27	A	B	C	D
4	A	B	C	D	16	A	B	C	D	28	A	B	C	D
5	A	B	C	D	17	A	B	C	D	29	A	B	C	D
6	A	B	C	D	18	A	B	C	D	30	A	B	C	D
7	A	B	C	D	19	A	B	C	D	31	A	B	C	D
8	A	B	C	D	20	A	B	C	D	32	A	B	C	D
9	A	B	C	D	21	A	B	C	D	33	A	B	C	D
10	A	B	C	D	22	A	B	C	D	34	A	B	C	D
11	A	B	C	D	23	A	B	C	D	35	A	B	C	D
12	A	B	C	D	24	A	B	C	D					

1 Which diagram shows two magnets repelling each other?

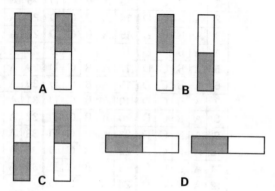

2 Which is a correct description of the forces between two charged objects?
- **A.** A positively charged object attracts a negatively charged object
- **B.** A positively charged object repels a negatively charged object
- **C.** Two negatively charged objects attract each other
- **D.** Two positively charged objects attract each other

3 A box is lifted from the floor and placed on a table.
Which statement is correct?
- **A.** The box has more gravitational potential energy when it is on the floor
- **B.** The box has more gravitational potential energy when it is on the table
- **C.** The box has more kinetic energy when it is on the floor
- **D.** The box has more kinetic energy when it is on the table

4 A car travels at constant speed on a level road. Which diagram shows the horizontal forces acting on the car?

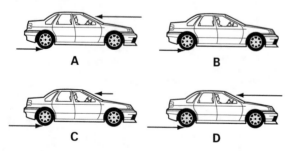

5 Which of the following results in the pitch of a sound being increased?
- **A.** Decreasing the amplitude
- **B.** Decreasing the frequency
- **C.** Increasing the amplitude
- **D.** Increasing the frequency

6 Which diagram shows the magnetic field pattern around a bar magnet?

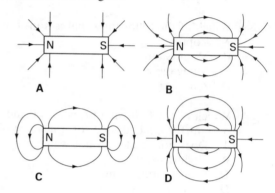

7 Choose the best description of a transformer.
- **A.** A device that changes the size of an alternating voltage
- **B.** A device that changes the size of a direct voltage
- **C.** A device that makes a voltage bigger
- **D.** A device that makes a voltage smaller

8 The diagram shows an electric bell.

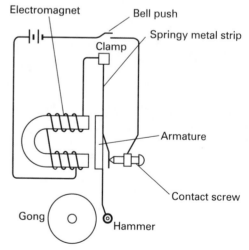

The purpose of the electromagnet is to
- **A.** attract the armature
- **B.** attract the contact screw
- **C.** attract the gong
- **D.** attract the hammer

9 When hair is combed using a nylon comb, the hair becomes positively charged and the comb becomes negatively charged.
Which could be a correct explanation of the charging of hair by a nylon comb?
A. Positively charged particles move from the hair to the comb
B. Positively charged particles move from the comb to the hair
C. Negatively charged particles move from the hair to the comb
D. Negatively charged particles move from the comb to the hair

10 When aircraft are being refuelled, an electrical connection is made to the Earth.
The purpose of this is
A. to create a high voltage on the aircraft
B. to discharge the aircraft battery
C. to prevent a high voltage on the aircraft
D. to recharge the aircraft battery

11 When an electric current passes in a metal, the current is due to
A. a flow of negatively charged electrons
B. a flow of negatively charged protons
C. a flow of positively charged electrons
D. a flow of positively charged protons

12 A force of 60 N is used to push a supermarket trolley for a distance of 8 m.
The amount of work done by the force is
A. 7.5 J
B. 7.5 N
C. 480 J
D. 480 N

13 A car travels at a steady speed on a level road. The driver notices a hazard and brakes.
The braking distance depends on
A. the chemical energy of the car
B. the combined gravitational potential energy and kinetic energy of the car
C. the gravitational potential energy of the car
D. the kinetic energy of the car

14 A runner does 4500 J of work in 10 s.
The power of the runner is
A. 450 W
B. 450 s/J
C. 4500 W
D. 4500 s/J

15 Which diagram shows the forces acting on a parachutist falling at terminal velocity?

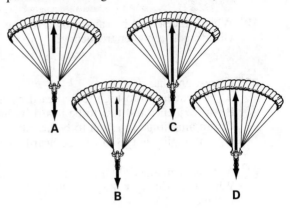

16 The point of a drawing pin has a small surface area. This is so that it
A. exerts a large force
B. exerts a large pressure
C. exerts a small force
D. exerts a small pressure

17 A person weighs 600 N. He stands on two skis, each of which has an area of 0.3 m^2.
The pressure on the ground is
A. 180 Pa
B. 360 Pa
C. 1000 Pa
D. 2000 Pa

Questions 18–19
A vibrating tuning fork causes the following trace on the screen of a CRO.

Here are four more CRO traces. The CRO controls have not been changed.

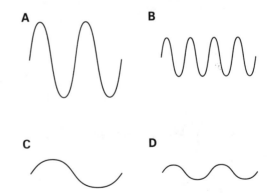

18 Which trace could be caused by the same tuning fork making a louder sound?

19 Which trace could be caused by a tuning fork vibrating at a higher frequency?

20 Ultrasound has
A. a frequency too high to be heard by humans
B. a frequency too low to be heard by humans
C. an amplitude too high to be heard by humans
D. an amplitude too low to be heard by humans

H21 A transformer has 200 turns on the secondary coil. When the primary voltage is 240 V, the secondary voltage is 12 V.
The number of turns on the primary coil is
A. 10
B. 200
C. 2400
D. 4000

H22 15 C of charge flow through the filament of a 240 V lamp in one minute.
The current in the lamp filament is
A. 0.25 A
B. 15 A
C. 16 A
D. 900 A

H23 An 800 kg car is travelling at 25 m/s.
Its kinetic energy is
A. 32 J
B. 10 000 J
C. 20 000 J
D. 250 000 J

Use the equation $KE = \frac{1}{2}mv^2$ to calculate the kinetic energy of the car.

H24 An 800 kg car increases its speed from 5 m/s to 17 m/s in 24 s.
The size of the unbalanced force required to do this is
A. 200 N
B. 400 N
C. 800 N
D. 1600 N

1	A	B	C	D	9	A	B	C	D	17	A	B	C	D
2	A	B	C	D	10	A	B	C	D	18	A	B	C	D
3	A	B	C	D	11	A	B	C	D	19	A	B	C	D
4	A	B	C	D	12	A	B	C	D	20	A	B	C	D
5	A	B	C	D	13	A	B	C	D	21	A	B	C	D
6	A	B	C	D	14	A	B	C	D	22	A	B	C	D
7	A	B	C	D	15	A	B	C	D	23	A	B	C	D
8	A	B	C	D	16	A	B	C	D	24	A	B	C	D

Universal changes

1 Which gas is present in the largest amounts in air?
A. Argon
B. Carbon dioxide
C. Nitrogen
D. Oxygen

2 Which process reduces the amount of carbon dioxide in the air?
A. Combustion
B. Photosynthesis
C. Respiration
D. Rusting

3 Increasing concentrations of carbon dioxide may
A. cause another 'ice age'
B. destroy the ozone layer
C. destroy rain forests
D. increase global warming

4 What is the approximate percentage of oxygen in the air?
A. 10%   B. 20%   C. 80%   D. 90%

5 Sulphur dioxide in the atmosphere millions of years ago came from
A. combustion of plants and animals
B. decomposition of plants and animals
C. gases escaping from volcanoes
D. outer space

6 Which of the following is an example of a longitudinal wave?
A. Light
B. Radio waves
C. Sound
D. Surface water waves

7 Which is the outer layer of the Earth?
A. The crust
B. The inner core
C. The mantle
D. The outer core

8 For light to be totally internally reflected at the boundary of a glass fibre, the angle of incidence must be
A. Greater than the angle of refraction
B. Greater than the critical angle
C. Less than the angle of refraction
D. Less than the critical angle

9 Which statement about transverse waves is true?
A. They all travel at the same speed in a vacuum
B. They all need a material to transmit them
C. They all have vibrations perpendicular to the direction of travel
D. They all have vibrations parallel to the direction of travel

H10 At the end of its main sequence, our Sun will expand to form
A. a blue dwarf
B. a blue giant
C. a red dwarf
D. a red giant

H11 A sound wave has a frequency of 240 Hz and a wavelength of 1.25 m.
The speed of the sound is
A. 192 m/s
B. 200 m/s
C. 300 m/s
D. 330 m/s

H12 On the surface of a pond, waves travel at 3.0 m/s. What is the wavelength of water waves with a frequency of 0.5 Hz?
A. 1.5 m
B. 2.5 m
C. 3.5 m
D. 6.0 m

Questions 13–14
Four words used to describe rocks are

1. crystalline
2. non-crystalline
3. hard
4. soft

13 Which two words describe granite?
A. 1 and 3
B. 1 and 4
C. 2 and 3
D. 2 and 4

14 Which two words describe sandstone?
A. 1 and 3
B. 1 and 4
C. 2 and 3
D. 2 and 4

15 High temperatures and high pressures turn
A. metamorphic rocks into igneous rocks
B. metamorphic rocks into sedimentary rocks
C. sedimentary rocks into igneous rocks
D. sedimentary rocks into metamorphic rocks

16 Four steps in the formation of sedimentary rocks are

1. burial
2. compression
3. deposition
4. weathering

The correct order of these steps is
A. 2, 1, 4, 3
B. 3, 4, 2, 1
C. 4, 2, 1, 3
D. 4, 3, 1, 2

17 In which rocks may fossils be found?
A. Igneous and metamorphic
B. Igneous and sedimentary
C. Metamorphic and sedimentary
D. Only igneous

18 When acid is added to limestone, the mixture fizzes. The gas given off is
A. carbon dioxide
B. hydrogen
C. nitrogen
D. oxygen

19 The drawing shows a sedimentary rock.

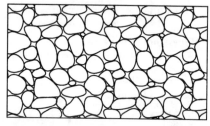

Which of the following statements is true?
A. Large crystals show the rock cooled quickly
B. Large crystals show the rock cooled slowly
C. Water has removed sharp edges from the large sediments in the rock
D. A high pressure was needed to form this rock

20 The Andes mountain range in South America was formed when
A. there was a series of volcanic eruptions
B. two plates moved apart
C. two plates moved past each other
D. two plates moved together

21 Rocks are returned to the mantle at
A. a constructive plate margin
B. a destructive plate margin
C. an earthquake
D. a volcanic eruption

22 The innermost part of the Earth is called the
A. core
B. crust
C. mantle
D. plates

23 Which metal is extracted from its ore by electrolysis?
A. Copper
B. Gold
C. Sodium
D. Zinc

24 An ore is a
A. mineral or mixture of minerals
B. mixture of metals
C. pure compound
D. pure metal

25 What is the name of an ore of iron?
A. Bauxite
B. Haematite
C. Rust
D. Slag

26 Pure copper is made from copper(II) sulphate solution by
A. distillation
B. electrolysis
C. evaporation
D. filtering

27 Which row in the table gives a correct use for cast iron and steel?

	Cast iron	Steel
A	car engines	car bodies
B	car bodies	car engines
C	spanner	car springs
D	car springs	spanner

H28 In the extraction of aluminium, aluminium oxide is dissolved in molten cryolite.
This mixture is the
A. anode
B. cathode
C. electrode
D. electrolyte

29 What is added, with iron ore and coke, to the blast furnace?
A. Calcium oxide
B. Limestone
C. Sand
D. Slag

30 An aluminium electrolysis factory is often sited close to a hydroelectric power station because
A. aluminium is needed by the hydroelectric power station
B. electricity is the major cost
C. land will be cheap for the factory
D. the process needs plenty of water

31 Which of the following will make existing metal stocks last longer?

1. Metals become cheaper.
2. Stopping recycling of metals.
3. New alternatives to metals being found.

A. 1 and 2
B. 2 and 3
C. 2 only
D. 3 only

32 In the blast furnace, iron oxide is reduced to iron. The reducing agent is
A. carbon dioxide
B. carbon monoxide
C. iron oxide
D. limestone

33 Which one of the following is **not** a pure metal?
A. Copper
B. Iron
C. Steel
D. Zinc

34 Which of the following is **not** a mineral?
A. Calcium carbonate
B. Gold
C. Iron sulphide
D. Water

35 Aluminium is extracted from bauxite by
A. electrolysis in molten cryolite
B. electrolysis of a solution
C. heating bauxite
D. reduction in a blast furnace

1	A	B	C	D	13	A	B	C	D	25	A	B	C	D
2	A	B	C	D	14	A	B	C	D	26	A	B	C	D
3	A	B	C	D	15	A	B	C	D	27	A	B	C	D
4	A	B	C	D	16	A	B	C	D	28	A	B	C	D
5	A	B	C	D	17	A	B	C	D	29	A	B	C	D
6	A	B	C	D	18	A	B	C	D	30	A	B	C	D
7	A	B	C	D	19	A	B	C	D	31	A	B	C	D
8	A	B	C	D	20	A	B	C	D	32	A	B	C	D
9	A	B	C	D	21	A	B	C	D	33	A	B	C	D
10	A	B	C	D	22	A	B	C	D	34	A	B	C	D
11	A	B	C	D	23	A	B	C	D	35	A	B	C	D
12	A	B	C	D	24	A	B	C	D					

Glossary

A

Active transport The movement of a substance against a concentration gradient, usually requiring energy.

Adaptation A feature that fits an organism to its environment.

Addiction Not being able to do without a drug such as alcohol, tobacco or heroin.

Absorber Dark-coloured objects are good absorbers of infra-red radiation. Light-coloured and silvered objects are poor absorbers.

Acceleration A change of speed or velocity. Defined as *increase in velocity/time taken* and measured in units of m/s².

Acid A substance that dissolves in water to form a solution with a pH below 7. An acid contains hydrogen which can be replaced by a metal to form a salt.

Activation energy The energy required to start a reaction. A **catalyst** lowers the activation energy.

ADH A hormone, produced by a gland in the brain, that prevents excessive excretion of water by the kidneys.

Aerobic Such a biological process needs oxygen to take place.

Air resistance A force that opposes motion through air. It acts in the opposite direction to the direction of motion and increases as the speed of the moving object increases.

Air Air is a mixture of gases. Approximately one-fifth is oxygen and four-fifths nitrogen.

Alcohol-emulsion test A test for fats. The substance is shaken with ethanol and some of the ethanol is then poured into water. A milky appearance indicates the presence of fat.

Alkali metal A metal in group I of the Periodic Table, e.g. sodium, potassium.

Alkali A metal oxide (base) or hydroxide that dissolves in water to form a solution with a pH greater than 7. An alkali is neutralised by an acid to form a salt and water.

Alkanes A family of hydrocarbons with a general formula of C_nH_{2n+2}. The simplest alkane is methane, CH_4.

Alkenes A family of hydrocarbons with a general formula of C_nH_{2n}. The simplest alkene is ethene, C_2H_4.

Allele Many genes have two or more forms, each called alleles.

Alloy A mixture of metals or a metal and carbon. Examples are brass and steel.

Alpha A type of nuclear radiation consisting of two neutrons and two protons.

Alternating current A current that changes direction periodically.

Aluminium A dull, silvery metal widely used especially in alloys.

Aluminium oxide Aluminium oxide (or alumina) is made from bauxite (aluminium ore). It is used for extracting aluminium.

Alveolus (plural: alveoli) A microscopic air sac in the lungs.

Amino acids The monomers from which proteins are made.

Ammeter An instrument used to measure the size of an electric current.

Ammonia A colourless gas which turns red litmus blue. Formula NH_3.

Amp The unit of electric current.

Amplitude The amplitude of a wave is the maximum displacement from the mean or rest position.

Amylase An enzyme which splits up carbohydrates into sugars.

Anaerobic Such a biological process can take place in the absence of oxygen.

Anchorage Secure fixing.

Angle of incidence When light strikes a boundary between two materials, this is the angle between the direction of travel of incident light and a line drawn at right angles to the boundary line.

Angle of reflection When light strikes a boundary between two materials, this is the angle between the direction of travel of reflected light and a line drawn at right angles to the boundary line.

Angle of refraction When light strikes a boundary between two materials, this is the angle between the direction of travel of refracted light and a line drawn at right angles to the boundary line.

Anode The positively charged electrode in electrolysis.

Antibody A protein produced by lymphocytes which helps to neutralise the effects of foreign cells or substances in the body.

Anus The end of the digestive system, where faeces leave the body.

Aorta The main artery of the body.

Armature The part of an electromagnetic device that moves. In a simple motor the armature is the part that includes the wire coil.

Artery A blood vessel that carries blood away from the heart.

Artificial selection Selection in which humans choose desirable characteristics to breed in other animals and plants.

Asexual A method of reproduction which does not involve the fusion of gametes.

Assortment The mixing up of genes during sexual reproduction.

Asteroid A piece of rock that orbits the Earth. The orbits of asteroids are between those of Mars and Jupiter.

Atmosphere All the air above the surface of the Earth.

Atom The smallest particle of an element which can exist.

Atomic number The number of protons in the nucleus of an atom.

Atrium (plural: atria) Chambers in the heart that receive blood from veins.

Attract To pull together.

Axis The centre of rotation.

B

Background radiation The particles and waves due to radioactivity from buildings, the ground and the atmosphere.

Bacteria Microscopic organisms that have cells and genetic material, but no nuclei.

Balanced forces Two forces are balanced if they are equal in size and opposite in direction.

Battery A number of electric cells connected together.

Bauxite The common ore of aluminium containing aluminium oxide with impurities such as iron(III) oxide.

Becquerel A unit of radioactivity, equal to a count rate of 1 count/s.

Benedict's test A test for reducing sugars. Benedict's reagent gives a brick-red precipitate when boiled with the reducing sugar.

Beta A type of nuclear radiation consisting of fast-moving electrons emitted from nuclei when they undergo radioactive decay.

Big Bang theory A theory that the whole Universe started with an enormous explosion.

Bile An alkaline liquid that emulsifies fats. It is produced by the liver.

Bitumen A thick black residue from fractional distillation of crude oil used for resurfacing roads.

Biuret test A test for protein, giving a lilac colour if protein is present.

Bladder An organ in the abdomen that stores urine.

Blast furnace A furnace used for extracting metals such as iron. Blasts of hot air are blown through the furnace.

Blind spot The part of the retina where the optic nerve leaves the eye.

Bloodstream The body system whose main function is to transport materials.

Boiling point A liquid turns rapidly to its vapour at a fixed temperature called the boiling point, which varies with pressure. The lower the pressure the lower the boiling point.

Bond breaking Breaking chemical bonds requires energy.

Bond making Forming chemical bonds releases energy.

Bonding The forces joining atoms together.

Bowman's capsule Part of a kidney tubule which collects the filtrate when blood is filtered.

Brain The part of the nervous system that co-ordinates most of the body's activities.

Braking distance The distance that a vehicle travels from when the brakes are applied until it stops.

Breeding Producing offspring.

Bronchiole The smallest tubes in the breathing system, ending in alveoli.

Bronchus (plural: bronchi) A main branch of the windpipe.

Burning (or combustion) The combination of a substance with oxygen to produce energy.

C

Calcium carbonate A compound present in a wide range of rocks including chalk, limestone and marble.

Camouflage Blending in with the surroundings.

Cancer A group of cells dividing much more rapidly than normal.

Canine A pointed tooth used for tearing and piercing.

Capillary A microscopic blood vessel that exchanges materials with body tissues.

Carbohydrate A food material used mainly as an energy source.

Carbon A non-metallic element.

Carbon dioxide A colourless gas produced when carbon or carbon compounds burn in a plentiful supply of oxygen. It is also formed when an acid acts on a carbonate.

Carbon monoxide A poisonous gas produced when carbon and carbon compounds burn in a limited supply of oxygen. Carbon monoxide is a good reducing agent.

Carnivore An animal that eats other animals.

Catalyst A substance which alters the rate of a reaction without being used up.

Cathode A negatively charged electrode in electrolysis.

Cell membrane The part of a cell that controls the entry and exit of materials.

Cell sap The liquid that fills the vacuole in a plant cell.

Cell wall The outer part of a plant cell, giving it shape and strength.

Cell In electricity, a cell is a combination of metals and chemicals that produces a voltage and can cause a current.

Cellulose A carbohydrate in a plant cell wall, which strengthens the wall.

CFC A type of compound formerly used as a propellant in aerosols. CFCs are now banned because they are destroying the ozone layer in the atmosphere.

Characteristic An observable feature (of an organism).

Charge A property of atomic particles. Electrons and protons have opposite charges and attract each other. The charge on an electron is negative and that on a proton is positive.

Chloride The ion formed when a chlorine atom gains an electron.

Chlorine A reactive greenish-yellow gas of the halogen family (group 7).

Chlorophyll A green pigment in plant cells that captures light energy for photosynthesis.

Chloroplast Part of a plant cell that contains chlorophyll.

Chromatography A way of separating mixtures, especially coloured substances, by letting them spread across a filter paper or through a powder. Each component in the mixture spreads at a different speed.

Chromosome Part of a cell that carries genetic information.

Ciliary muscle A muscle in the eye that can alter the shape of the lens.

Circuit breaker A device used in place of a fuse to disconnect the supply voltage to a circuit if a fault develops.

Clone A group of genetically identical cells or organisms.

Clot A mixture of protein fibres and blood cells formed to stop bleeding.

Coal A fossil fuel formed from decayed tree-like plants.

Coke An impure form of carbon made from coal.

Collecting duct The final part of a kidney tubule, where the concentration of the urine is adjusted.

Combination The joining together (or **combining**) of atoms of different elements to form a compound.

Combustion See **burning**.

Comet An object made up of ice and rock, that orbits the Sun. The orbit of a comet is elliptical and can be in any direction and any plane.

Commutator A device that makes the electrical connection to the moving coil of a motor or generator. It allows current to pass in or out of the coil while it rotates.

Competition A struggle between organisms when there is not enough of a resource to satisfy the needs of all of them.

Component A device in a circuit that transfers energy from electricity into a useful form.

Compound A substance formed by joining atoms of different elements together.

Compression In a sound wave, a compression is a region in the material that transmits the wave where the particles are closer together than normal.

Concentration gradient A difference in concentration of a substance between two regions.

Condensing Changing a vapour (or gas) into a liquid. This change is accompanied by a giving out of energy.

Conductor A material that allows heat or electricity to pass through it.

Cone A sensory cell in the retina of the eye, responsible for colour vision in bright light.

Consolidation One of the stages in forming a sedimentary rock.

Constrict To get narrower.

Constructive boundary Where, when two plates are moving apart, hot molten rock comes to the surface and forms new rocks.

Consumer An organism that eats another organism.

Contact screw Part of the make-and-break circuit of an electric bell.

Continuous variation Variation where a feature cannot be divided into distinct groups, e.g. human height and human mass.

Contract To get shorter.

Convection Movement of a fluid (gas or liquid) due to differences in density. It is caused by one region of the fluid being hotter or colder than the surrounding fluid.

Copper A reddish-brown shiny transition metal.

Core 1. In electromagnetism, the material inside the coils of a transformer or electromagnet.
2. The dense material at the centre of the Earth. It is composed of a solid inner core and a liquid outer core.

Cornea The transparent part of the front of the eye, largely responsible for focusing light on the retina.

Corrosion The wearing away of the surface of a metal by chemical reactions with oxygen and water. Rusting is an example of corrosion.

Coulomb The unit of charge. One coulomb is the charge that flows past a point when a current of 1 amp passes for 1 second.

Covalent A type of bonding involving the sharing of one or more pairs of electrons. The electrons are given by the atoms combining.

Cracking The breaking down of long-chain hydrocarbon molecules with heat or catalyst to produce small molecules useful for making polymers.

Critical angle Applies when light strikes a boundary between two materials and the speed of light in the material that it is travelling in is less than the speed of light in the material beyond the boundary. If the angle of incidence at the boundary is greater than the critical angle all the light is reflected.

Crude oil (or petroleum) A complicated mixture of hydrocarbons produced by the action of high temperature and high pressure on the remains of sea creatures in the absence of air. It is trapped between impermeable rocks.

Crust The outer layer of the Earth.

Cryolite A sodium aluminium fluoride used as a solvent for aluminium oxide in the extraction of aluminium.

Crystal A piece of solid substance that has a regular shape. The regular shape is caused by a regular arrangement of particles in the crystal.

Crystallisation A process producing crystals. Crystals are formed when a molten substance is cooled or when a hot, saturated solution is cooled. Slow crystallisation produces large crystals and rapid crystallisation produces small crystals.

Cubic A common shape for crystals, e.g. salt, magnesium oxide.

Current A flow of charge.

Cuticle An outer skin.

Cutting A part of a shoot, cut off and planted to produce a new plant.

Cycle In wave motion, one cycle is a trough and a crest for a transverse wave, or a compression and a rarefaction for a longitudinal wave.

Cytoplasm The part of a cell where most chemical reactions occur.

D

Darwin The scientist who proposed the theory of evolution by natural selection.

Decay The breakdown of dead and waste matter by bacteria and fungi.

Denatured When enzymes are heated above about 40 °C the active

sites are destroyed and the catalytic activity is destroyed.

Density The mass of a given volume of a substance. It has units of kg/m^3 or g/cm^3. When the density is high the particles are closely packed.

Deposition Formation of a **deposit**, i.e. a layer of precipitate.

Depressant A drug that slows down reactions, e.g. alcohol.

Destructive boundary A boundary where a denser oceanic plate moves under a continental crust plate. Rocks are returned to the mantle for recycling.

Detergent A cleaning agent. There are two main types: soaps and soapless detergents.

Detritus feeder An animal that feeds on dead and waste matter.

Diabetes A disease, caused by insufficient insulin, that results in high blood sugar levels.

Dialysis Using a partially permeable membrane to adjust the composition of a liquid mixture.

Diaphragm A sheet of muscle that separates the thorax from the abdomen.

1,2-Dibromoethane The addition product formed when bromine reacts with ethene.

Diesel oil One of the fractions produced on fractional distillation of crude oil.

Diffraction The spreading out of waves when they pass through a gap or around the edge of an obstacle.

Diffusion The spreading out of a substance, due to the kinetic energy of its particles, to fill all of the available space.

Digestion The breakdown of large, insoluble molecules into smaller, soluble molecules.

Dilate To get wider.

Diode A device in electronics that allows current to pass only in one direction.

Direct current Current that does not change direction.

Discontinuous variation Variation where a feature can be divided into distinct groups, e.g. human blood groups.

Dispersion The splitting of light into its constituent colours.

Displacement The distance and direction that an object has moved from its starting point or original position.

Displacement reaction A reaction in which one metal replaces another, e.g. copper(II) sulphate + iron → copper + iron(II) sulphate.

Distillation A process of purification involving boiling followed by condensation.

DNA The chemical that carries the genetic code for characteristics.

Dominant The type of allele that controls the development of a characteristic when it is present on only one of the chromosomes.

Double circulation Circulation in which blood flows from the heart to the lungs, then back to the heart before being pumped to the rest of the organs.

E

Ear Part of the body that contains receptors for sound and balance.

Earth In electricity, a safety wire that provides a connection to the Earth's surface.

Echo A reflection of a sound or ultrasound.

Effector A muscle or gland which brings about the response to a stimulus.

Efficiency The proportion of energy input transferred to a useful output.

Elastic A material is elastic if it returns to its original shape when a force is removed.

Elastic limit The maximum force that can be applied to a material for it to remain elastic.

Electricity A general term that describes the presence of a voltage or current.

Electrolysis The splitting up of an electrolyte, either molten or in aqueous solution, by electricity.

Electrolyte A chemical compound which, in aqueous solution or when molten, conducts electricity and is split up by it. Acids, bases, alkalis and salts are electrolytes.

Electromagnet A device that consists of a coil of wire, usually wound on a soft iron core. It attracts magnetic objects when a current passes through the coil.

Electromagnetic induction A voltage in a conductor caused by a changing magnetic field or movement through a magnetic field.

Electromagnetic radiation This travels as a transverse wave. It includes radio waves, microwaves, infra-red, light, ultraviolet, X-rays and gamma rays.

Electron A fundamental negatively charged particle that orbits the nucleus of an atom and is responsible for electrical conduction in metals.

Electrostatic forces Forces caused by charges. Like charges repel and unlike charges attract.

Element A pure substance that cannot be split up into anything simpler.

Embryo A developing organism before birth.

Emitter Dark-coloured objects are good emitters of infra-red radiation. Light-coloured and silvered objects are poor emitters.

Emulsifies Breaks down large drops into smaller droplets.

Endothermic reaction A reaction in which energy is taken in from the surroundings.

Energy The ability to do work, e.g. cause motion.

Energy change The difference between the energy in the products of a chemical reaction and the reactants.

Energy level diagram A diagram showing the energy content at stages during a reaction.

Environmental Concerned with the surroundings.

Enzymes Proteins which act as biological catalysts.

Epidermis An outer layer of cells.

Equilibrium A **reversible reaction** is in equilibrium when the rate of the **forward reaction** equals the rate of the **reverse reaction**. If conditions change, the equilibrium may move to the right to produce more products or move to the left to produce more reactants.

Erode Wear away.

Erosion The process in which rocks are worn away.

Ethene The simplest **alkene** with a formula C_2H_4.

Eutrophication The result of the introduction of excessive amounts of nutrients, often nitrates and phosphates, into rivers or lakes.

Evaporating The process by which a liquid changes to a vapour, due to particles leaving the surface of the liquid. This happens at temperatures below the boiling point but is fastest when the liquid is boiling.

Excretion The getting rid of waste materials from the body.

Exothermic reaction A reaction in which energy is lost to the surroundings.

Expand To increase in size.

Explosion A very rapid reaction accompanied by a large expansion of gases.

Extrusive A rock of this type crystallises in the surface of the Earth, e.g. basalt.

Eye An organ containing receptors for light.

F

Faeces Mainly indigestible food at the end of digestion.

Fat A food material used mainly as an energy source. Fats are fomed

when fatty acids and glycerol combine.

Fatigue Tiredness.

Fatty acid One of the chemicals which make up a fat.

Fault Breaks in the ground where plates join, e.g. San Andreas Fault in California.

Fermentation The process in which enzymes in yeast change glucose into ethanol and carbon dioxide.

Fertilisation The fusion (joining together) of two gametes (sex cells).

Fertiliser A substance added to the soil to improve the growth of plants.

Fertility How able an individual is to produce offspring.

Filtering (or filtration) A method of separating a solid from a liquid.

Force A force is a push or a pull. All forces can be described as *object A pulls/pushes object B*.

Formula mass Mass in grams of 1 mole of material, e.g. formula mass of carbon dioxide is 44 g.

Fossil The remains of plant or animal bodies which have not decayed and disappeared but have been preserved. Fossils may be found in sedimentary and metamorphic but not igneous rocks.

Fossil fuels Fuels such as coal, oil and natural gas produced in the Earth over long periods of time.

Fovea The part of the retina which contains the highest density of cones.

Fractional distillation A method of separating liquids with different boiling points.

Free electron In a metal, an electron that can move around within the metal and does not stay in orbit around a nucleus.

Freezing Changing from a liquid to a solid at the freezing point.

Frequency The frequency of a wave is the number of cycles that occur each second. It is measured in hertz (Hz).

Friction A force that opposes two surfaces slipping or sliding over each other.

Fruit A structure formed by a flower after fertilisation.

Fuel A substance which burns to produce energy.

Fuse In electricity, a safety device fitted to plugs. It consists of a wire that melts and breaks the circuit if the current is too high.

Fusion The joining together of atomic nuclei. When small nuclei fuse together energy is released but the fusion of large nuclei absorbs energy. The energy radiated by the Sun comes from nuclear fusion.

G

Galaxy A collection of stars held together by gravitational forces.

Gall bladder An organ attached to the liver. It stores bile.

Gamete A sex cell that fuses with another sex cell.

Gamma ray Very short wavelength electromagnetic radiation given off from an atomic nucleus.

Gas A state of matter in which the particles are widely spaced.

Geiger–Müller tube A device that detects nuclear radiation.

Gene A section of a DNA molecule that controls the development of a characteristic.

Generator This produces electricity when a magnet spins inside a coil of wire or a coil of wire spins inside a magnetic field.

Genetic Concerned with inheritance.

Genotype The genetic make-up of an individual comprising all the alleles.

Geosynchronous A term used to describe a satellite that has a period of rotation around the Earth of 24 hours. It stays above the same point on the Earth's surface.

Geothermal A type of energy source involving the extraction of heat from the Earth's crust.

Geotropism The response of a plant organ to the stimulus of gravity.

Giant structure A crystal structure in which all of the particles are linked together in a network of bonds extending throughout the crystal, e.g. diamond.

Gland Part of an organism that produces materials for use in another part of the organism.

Glomerulus The part of the kidney where blood is filtered, consisting of a knot of blood vessels.

Glucose A simple carbohydrate; a monomer of polymers such as starch and cellulose.

Glycerol One of the chemicals which combine to make up a fat.

Glycogen A carbohydrate, stored in animals mainly in the liver and muscles.

Gradient The slope of a graph.

Gravitational force The attractive force that exists between all objects.

Gravitational potential energy The energy transfer caused by a change in height above the surface of the Earth or other planet.

Greenhouse effect The heating of the Earth caused by the increase in concentration of atmospheric carbon dioxide and other 'greenhouse gases'. It results in **global warming**.

Group A vertical column in the **Periodic Table**.

Growth An increase in the size of an organism.

Guard cells Cells that control the width of stomata (tiny holes in the surfaces of leaves).

H

Haematite One of the common iron ores containing iron(III) oxide.

Haemoglobin A red pigment in red blood cells, responsible for most oxygen transport.

Half-life In radioactivity, the average time it takes for half the radioactive nuclei present in a substance to decay.

Halogen An element in group 7 of the Periodic Table. The word 'halogen' means 'salt-producer'.

Helium A noble gas which is used in balloons because it has a low density.

Herbivore An animal that eats only plants.

Hertz The unit of frequency.

Heterozygous Containing both the dominant and the recessive allele of a gene.

Homeostasis Keeping conditions inside the body constant, e.g. keeping body temperature and blood sugar levels constant.

Homozygous Containing either two dominant alleles of a gene or two recessive alleles.

Hooke's law states that the extension of a material is proportional to the force acting on it.

Hormone A chemical, produced by one part of an organism, that controls a process in another part of the organism.

Hydraulic Using a liquid as a method of transmitting pressure. It allows forces to be **magnified**.

Hydroelectric A method of generating electricity using moving water. The water can be from a fast-moving river or from a dam or reservoir.

Hydrogen The gas with the lowest density. It burns with a squeaky pop when a lighted splint is put into it.

Hydrogen chloride A colourless gas which dissolves in water to form hydrochloric **acid.**

Hydroxide An ion present in alkalis.

I

Igneous rocks Rocks that have cooled and solidified as crystals from molten rock, e.g. granite.

Immune Not affected by an infectious microbe.

Impermeable rocks Rocks which do not allow water or gas to pass through them.

Impulse The form in which information is transmitted in the nervous system.

Incisor A chisel-shaped tooth at the front of the mouth used for cutting.

Inert Unreactive.

Infection Multiplication of a disease-producing microbe inside the body.

Infrared A type of electromagnetic radiation with a wavelength longer than that of light.

Ingest Take food into the body.

Inhibitor A catalyst that slows down the rate of a reaction.

Insoluble An insoluble substance will not dissolve.

Insulator A thermal insulator reduces the energy flow between a hot or cold object and its surroundings. An electrical insulator does not allow current to pass in it.

Insulin A hormone, produced by the pancreas, that lowers blood sugar levels.

Intercostal muscles Muscles between the ribs that contract to move the rib cage.

Intrusion A body of molten rock which forces its way between layers of rock.

Intrusive rocks Igneous rocks which crystallise inside the Earth, e.g. granite.

Iodine A grey-black solid halogen element.

Ion A positively or negatively charged particle formed when an atom or group of atoms loses or gains electrons.

Ionic A type of bonding involving complete transfer of one or more electrons from a metal atom to a non-metal atom. Doing this forms **ions**.

Iris The coloured part of the eye that controls the amount of light entering the eye.

Iron A grey transition metal with strong magnetic properties.

Iron(III) oxide Compound of iron present in many iron ores.

Isotopes Atoms with the same atomic number, but different mass numbers.

J

Joule The unit of work and energy.

Jupiter The largest planet in the Solar System. The orbit of Jupiter is between the orbits of the asteroids and Saturn.

K

Kidney An organ that excretes waste materials and controls the water and salt content of the blood.

Kilowatt A unit of power equal to one thousand watts.

Kilowatt-hour The energy flow through a one-kilowatt appliance in one hour.

Kinetic energy The energy an object has due to its movement.

L

Lactic acid An acid produced when milk sours. It is also produced in anaerobic respiration.

Lamarck A scientist who proposed a theory of evolution based on the inheritance of acquired characteristics. This theory is now discounted.

Lamp A device that gives out light when an electric current passes through it.

Large intestine Part of the digestive system that absorbs water from the indigestible food.

Lattice Ionic bonding leads to a crystalline structure called a lattice.

Lead A dull dark grey soft metal with a high density.

Leaf Part of a plant whose main function is to photosynthesise (manufacture food).

Lens Part of the eye that helps to focus light on the retina.

Light The part of the electromagnetic spectrum that is detected by the eye.

Light-year An astronomical measurement equal to the distance that light travels in one year.

Limewater A saturated solution of calcium hydroxide. It turns milky when carbon dioxide passes through it.

Limit of proportionality The maximum force that can be applied to an object for the extension to be proportional to the applied force.

Limiting factor A factor such as light intensity that limits the rate of a reaction.

Lipid Another name for a fat; made up of fatty acids plus glycerol.

Lipase An enzyme that digests fat into fatty acids and glycerol.

Liquid A state of matter.

Live Energy flows through this conductor in the mains supply.

Liver An organ responsible for many processes including the production of bile, the breakdown of amino acids into urea and the storage of sugars as glycogen.

Longitudinal A type of wave motion in which the oscillations are parallel to the direction of wave motion.

Lung The organ where gases are exchanged between the blood and air.

M

Magma Semi-molten rock under the solid **crust** of the Earth.

Magnesium A reactive metal in Group 2 of the Periodic Table.

Magnet This attracts magnetic materials such as iron, steel and nickel. It can attract and repel other magnets.

Magnetic field The region around a magnet or electric current where there is a force on magnetic materials.

Mains electricity The 240 V supply used at home and at work.

Malleable Metals are malleable because they can be beaten into thin sheets.

Maltose A carbohydrate consisting of two glucose molecules combined together.

Manganese A transition metal.

Mantle The part of the Earth between the crust and the core.

Marble A metamorphic rock produced by the action of high temperatures and pressures on limestone.

Mass number The number of protons plus the number of neutrons in an atom.

Mass This is measured in kilograms. The mass of an object does not change.

Meiosis A type of cell division, often during the formation of gametes, in which the number of chromosomes in cells is halved.

Melting A solid changes to a liquid at the melting point.

Mercury 1. The innermost planet in the Solar System.
2. The only liquid element at room temperature.

Metamorphic A type of rock that was originally either igneous or sedimentary and has been altered by the effects of high temperatures and pressures, e.g. marble.

Metamorphosis The process producing **metamorphic** rocks.

Methane The simplest member of the alkane family with a formula CH_4. It is the major component of natural gas.

Microbe A microscopic organism.

Microwave Short-wavelength radio waves used for cooking food and telecommunications.

Mineral A naturally occurring substance of which rocks are made.

Mitochondria Parts of a cell where most of the reactions in respiration occur.

Mitosis A type of cell division, usually during the formation of body cells, where the cells formed have genetic information identical with that of the original cell.

Moh's scale A scale of hardness of materials. Diamond, with a value of 10, is the hardest substance.

Molar A tooth at the back of the mouth. In humans it has a flat shape for grinding.

Molar volume Volume occupied by 1 mole of a gas. At room temperature and atmospheric pressure the molar volume is 24 dm^3.

Mole The amount of substance containing 6×10^{23} particles.

Molecular A type of structure built up of molecules. A substance with a molecular structure has a low melting and boiling point.

Molecule The smallest part of an element or compound which can exist on its own.

Moment The effect of a force in causing turning or rotation.

Monomer A small molecule which joins together with other molecules to produce a **polymer**.

Moon The Earth's natural satellite.

Motor A device consisting of a coil of wire and some magnets. The coil rotates when a current passes through it.

Motor A type of neurone (nerve cell) that carries information from the brain or spinal cord to an effector.

Movement Change of position.

Mucus A slimy fluid.

Muscle An organ which brings about movement.

Muscular Concerned with muscles.

Mutation A spontaneous or induced change in genetic information.

N

Negative The sign of the charge on an electron.

Negative feedback A mechanism which restores the original condition, e.g. a rise in blood sugar level leads to the secretion of insulin which in turn brings about a reduction in blood sugar level.

Neon A noble gas in group 0 of the Periodic Table.

Neurone A nerve cell.

Neutral 1. Having no overall charge.
2. A neutral substance has a pH of 7.

Neutralisation A reaction in which an acid reacts with a base or alkali.

Neutron An uncharged nuclear particle similar in mass to a proton.

Nitric acid An acid produced by the oxidation of ammonia.

Nitrifying bacteria Soil bacteria that convert ammonium ions into nitrate ions.

Nitrogen The commonest gas in the atmosphere.

Nitrogen-fixing bacteria Bacteria that can convert gaseous nitrogen into nitrogen-containing compounds.

Noble gas An element in group 0 of the Periodic Table.

Noise Sound consisting of an irregular mixture of frequencies.

Non-metal Most of the elements are metals with characteristic properties, e.g. shiny, high density, good conductors of heat and electricity. Non-metals, e.g. carbon and sulphur, form acidic oxides, metals form alkaline or neutral oxides.

Non-renewable A type of energy source that will run out and cannot be replaced.

Nose The organ of smell, containing receptors that are sensitive to chemicals.

Nucleon A proton or neutron in a nucleus.

Nucleus 1. The central part of an atom.
2. The part of a cell which contains genetic information.

Nutrients Food materials.

Nutrition Obtaining food and absorbing useful substances from it.

O

Oesophagus The gullet, part of the digestive system leading from the mouth to the stomach.

Oestrogen A hormone, secreted by the ovaries, that helps to control reproductive processes.

Ohm The unit of electrical resistance.

Omnivore An animal that eats both animals and plants.

Open cast In open cast mining rocks are dug out of the ground by removing soil and surface rocks and digging out with an excavator.

Optic nerve A nerve leading from the eye to the brain.

Orbit The circular or elliptical path of one astronomical object around another.

Organ system A group of organs with a common function.

Organ A group of tissues with a common function.

Organic compounds Compounds of carbon, with other elements such as hydrogen, oxygen, nitrogen, etc. Many such compounds are present in living matter.

Oscillation A to-and-fro or side-to-side movement such as that of a particle transmitting a wave.

Osmosis The net movement of water molecules through a partially permeable membrane from a region of high concentration of water molecules to a region of lower concentration.

Ovary An organ in the female body that produces eggs and sex hormones.

Oxidation A reaction in which a substance gains oxygen, loses hydrogen or loses electrons. The opposite of oxidation is **reduction**.

Oxide A compound of an element with oxygen. A basic oxide is an oxide of a metal. A neutral oxide, e.g. carbon monoxide, has no reaction with acids or alkalis and has a pH of 7. Acidic oxides are non-metal oxides which react with alkalis to form a salt and water.

Oxidise Gain oxygen, lose hydrogen or lose electrons.

Oxygen debt The oxygen needed to oxidise lactic acid that builds up during anaerobic respiration.

Oxygen A reactive element in Group 6 of the Periodic Table.

P

Palisade mesophyll Cells under the upper epidermis of a leaf, where most photosynthesis occurs.

Pancreas An organ near the stomach which secretes both digestive enzymes and hormones concerned with the control of blood sugar.

Paraffin One of the fractions produced in the fractional distillation of crude oil.

Paraffin oil Oil under which alkali metals are stored.

Parallel A type of circuit containing more than one current path.

Partially permeable Allowing small molecules to pass through quickly, but larger molecules more slowly.

Period A horizontal row in the Periodic Table.

Peristalsis The movement of food through the digestive system, brought about by contraction of muscles in the walls or the digestive organs.

Pesticide A chemical that is designed to kill pests.

Petrol A low boiling point fraction produced from fractional distillation of crude oil.

Petroleum See crude oil.

Petroleum gas (or natural gas) Gas found with petroleum, made up largely of methane.

pH A scale measuring acidity and alkalinity.

Phenotype All the observable characteristics of an organism.

Phloem A plant tissue responsible largely for transporting food materials around the plant.

Phosphorus A solid non-metallic element. Phosphorus is an element required for good plant growth.

Photosynthesis A process taking place in green parts of a plant. Water and carbon dioxide react together in sunlight in the presence of chlorophyll to produce sugars and oxygen gas.

Pivot The **axis** or centre of rotation.

Planet A large object that orbits a Sun.

Plasma The liquid part of blood.

Plastic A material is plastic if it keeps its new shape after a deforming force is removed.

Platelets Cell fragments in blood which assist blood clotting at wounds.

Plates Huge sections of the Earth's crust which float on the mantle.

Pluto The outermost planet in the Solar System.

Poles The parts of a magnet where the magnetism is strongest.

Pollution Unwanted, harmful materials in the habitat.

Polymer A long-chain molecule built up of a large number of small units, called **monomers**, joined together by a process called **polymerisation**.

Positive The sign of the charge on a proton.

Potassium A reactive alkali metal in group I of the Periodic Table. Potassium is an element required for good plant growth.

Power The energy transfer per second. Measured in watts (W).

Precipitate An insoluble substance formed in a chemical reaction involving solutions.

Precipitation A reaction in which a precipitate is formed.

Predator An animal that catches and eats other living animals.

Premolar A tooth in front of the molars. In humans it has a flat shape for grinding.

Pressure The force acting per unit area. Measured in Pa or N/m^2.

Prey An animal that is captured for food by another animal.

Primary The input coil of a transformer.

Producer An organism, usually a plant, that can produce its own food.

Proportion Two quantities are in proportion if the effect of doubling one quantity is to double the other. A graph of two quantities in proportion is a straight line through the origin.

Protease An enzyme that breaks down proteins into amino acids.

Protein A food material needed mainly for growth and repair. A protein is a natural condensation polymer made up of amino acids.

Proton A positively charged nuclear particle.

Pulmonary artery The blood vessel that takes deoxygenated blood from the heart to the lungs.

Pulmonary vein The blood vessel that takes oxygenated blood from the lungs to the heart.

Pupil The hole surrounded by the iris in the eye.

R

Radiation 1. Nuclear radiation is emitted when an unstable nucleus decays.
2. Electromagnetic radiation is a wave that forms part of the electromagnetic spectrum.

Radio waves Long-wavelength electromagnetic waves used for communications.

Radioactive A radioactive substance has nuclei that are unstable. They change to more stable nuclei by emitting alpha particles, beta particles or gamma rays or a combination of these.

Radiocarbon dating A method of estimating the age of dead biological material by measuring the amount of carbon-14 present.

Random movement Particles are moving with no pattern in their motion.

Random With no set order or pattern. Describes the motion of the particles in a gas and the pattern of radioactive decay.

Rarefaction In a sound wave, a region in the material that transmits the wave where the particles are further apart than normal.

Reactants The substances which react together to form **products**.

Reaction time The time that it takes a person to react to an event.

Reactivity series A list of metals in order of reactivity with the most reactive metal at the top of the list.

Real image An image that can be formed on a screen.

Receptor A cell which is sensitive to a stimulus.

Recessive A type of allele which has to be present on both chromosomes to control a characteristic.

Red cells Blood cells containing haemoglobin whose function is to transport oxygen.

Red giant A phase in the life cycle of a star. It follows the main sequence and is caused by the star expanding and cooling.

Reducing agent A substance that brings about the reduction of another substance. Common reducing agents are hydrogen and carbon monoxide.

Reduction See **oxidation**.

Reflex action An automatic response to a stimulus.

Refraction The change in speed of a wave as it passes from one substance into another. It causes a change in wavelength and may cause a change in direction.

Renal artery The vessel that supplies blood to the kidneys.

Renal vein The vessel that takes blood away from the kidneys.

Renewable An energy source that will not run out.

Repel To push away.

Reproduction Producing offspring.

Resistance A measure of the opposition to electric current. Measured in ohms (Ω).

Resistive A general term to describe the forces acting in opposition to a moving object.

Resistor A circuit component with a fixed resistance.

Respiration A process taking place in cells where sugars are reacted with oxygen to produce carbon dioxide and water with the release of energy.

Respire To release energy from food materials.

Response Behaviour of an organism when stimulated.

Retina The inner layer of the eye, containing light-sensitive cells.

Reversible reaction A reaction which can go forwards or backwards depending upon conditions.

Rib A bone protecting organs in the thorax, also important in breathing.

Rib cage The collection of ribs.

Rod A light-sensitive cell in the retina, used in dim light.

Root hairs Cells in **roots** largely responsible for the uptake of water and mineral ions.

Root The organ of a plant that anchors it in the ground and takes in water and mineral ions.

Rust A complex hydrated iron oxide formed when iron or steel reacts with air and water.

S

Salivary gland A gland near the mouth that secretes saliva which contains carbohydrase enzyme.

Salt A substance which is formed as a product of neutralisation.

Satellite An object that orbits the Earth or other astronomical body.

Saturated compound A compound which contains only single covalent bonds, e.g. ethane C_2H_6. An **unsaturated** compound contains one or more double or single bonds.

Saturated solution A solution in which no more solute will dissolve, providing the temperature remains constant.

Sclera The outer, white part of the eye.

Secondary The output coil of a transformer.

Sedimentary Type of rock that is composed of compacted fragments of older rocks which have been deposited in layers on the floor of a lake or sea, e.g. sandstone.

Seismometer An instrument that detects Earth tremors and earthquakes.

Selective breeding Where humans choose desirable characteristics when breeding other organisms.

Sensitivity The ability to react to changes in the environment.

Sensory Able to detect changes in the environment.

Series A type of circuit that has only one current path.

Sexual Involving sex or gametes.

Shale A sedimentary rock made up of very fine mud or clay particles compressed to form the rock.

Shell Electrons are around the nucleus in shells or energy levels. Each shell can hold a maximum number of electrons.

Silicon(IV) oxide The chief chemical constituent of sand.

Silver chloride A white insoluble silver compound precipitated when a solution containing chloride ions is added to silver nitrate solution.

Skin An organ covering the outside of an organism.

Small intestine The part of the digestive system where digestion is completed and absorption of soluble foods occurs.

Sodium A reactive alkali metal.

Sodium hydroxide A cheap alkali produced by the electrolysis of brine.

Solar System Our Sun and all the planets, asteroids and comets that orbit it.

Soluble A soluble substance will dissolve.

Sound A longitudinal wave detected by the ear.

Spark Light that is emitted when electric charge passes through air.

Spectrum A group of waves arranged in order of wavelength or frequency.

Speed *Distance travelled ÷ time taken.* Measured in m/s.

Spinal cord An extension of the brain, mainly responsible for transmitting information to and from the body.

Spongy mesophyll A layer near the under surface of a leaf with large air spaces to allow rapid diffusion of gases.

Star An astronomical body like our Sun that emits electromagnetic radiation. Most stars emit light.

Starch A carbohydrate polymer stored by most plants.

Static electricity A build-up of electric charge on an object.

Steel An alloy of iron containing a small percentage of carbon.

Stem The organ in a plant that supports the leaves and transports materials.

Stimulus A change in the environment which can be detected by organisms.

Stoma (plural: stomata) A tiny hole in the surface of a leaf that allows gases to enter and exit.

Stomach An organ in the digestive system where digestion of protein begins.

Stopping distance The distance that a vehicle travels between the driver applying the brakes and the vehicle coming to rest.

Subduction Movement of one plate under another.

Substrate A liquid in which enzymes can operate.

Sugar A carbohydrate with small molecules.

Sulphur dioxide A pollutant gas released during the combustion of some fossil fuels.

Sulphuric acid Sulphuric acid is the acid produced by the Contact process. Salts made from it are called **sulphates**.

Sun A light-emitting star at the centre of a Solar System.

Supernova A star in a very bright phase prior to exploding.

Support Hold up.

Surface The edge or boundary of a substance.

Surface area All the outside of an organism.

Suspensory ligament A structure which holds the lens of the eye in place.

Sweat A liquid produced by glands in the skin. It cools the body as it evaporates.

Switch In a circuit, the switch makes electrical contact when it is closed and breaks contact when it is open.

Synapse The junction between two nerve cells.

Synthesis See **combination**.

Synthetic polymers Synthetic polymers are made from monomers obtained from products from crude oil.

T

Temperature How hot an object is. Measured in °C or K.

Terminal velocity The maximum speed of an object moving through a fluid. At terminal velocity the driving force and resistive forces are of equal size and act in opposite directions.

Testis An organ in males that produces sperm and sex hormones.

Testosterone A hormone produced in the testis which controls reproductive processes.

Thermal energy The energy that an object has due to its temperature.

Thermit reaction The thermit reaction is a displacement reaction in which iron is obtained from iron(III) oxide using aluminium powder. This is an exothermic reaction.

Thermoregulatory centre Part of the brain that co-ordinates the processes which control body temperature.

Thinking distance The distance that a vehicle travels during the driver's reaction time.

Tissue A group of similar cells with a common function.

Tissue culture A special medium in which clones of cells are produced.

Tissue fluid The part of blood plasma that has passed out of the capillaries into the tissues.

Total internal reflection This occurs when light strikes a boundary between two materials and the speed of light in the material that it is travelling in is less than the speed of light in the material beyond the boundary. If the angle of incidence at the boundary is greater than the critical angle all the light is reflected.

Trachea The windpipe that connects the lungs to the throat.

Transformer An electromagnetic device that changes the size of an alternating voltage.

Transition metals The block of metals between the two parts of the main block in the Periodic Table. Transition metals are usually dense metals that are much less reactive than alkali metals.

Transpiration The loss of water vapour from the shoot of a plant.

Transplant Transferring an organ from one person to another.

Transportation One of the steps in the formation of sedimentary rocks.

Transverse A type of wave motion in which the vibrations are at right angles to the direction of wave travel.

Trend A pattern in properties.

Turbine A device that drives a generator. The turbine is driven by moving water, high pressure steam or hot exhaust gases.

Turgor The pressure exerted by the contents of a plant cell on its cell wall.

U

Ultrasound A longitudinal wave like sound, but with a frequency too high to be detected by humans.

Ultraviolet Electromagnetic radiation with a wavelength shorter than that of light.

Universal indicator A mixture of indicators used for finding the pH of a substance.

Universe Everything that exists.

Urea A waste material, produced in the liver from the breakdown of amino acids, excreted by the kidneys in urine.

Ureter A tube carrying urine from a kidney to the bladder.

Urethra A tube carrying urine from the bladder to the outside.

Urine A waste fluid, produced by the kidneys, containing urea, excess water and excess salts.

V

Vacuole A fluid-filled sac in most plant cells.

Vacuum A region containing nothing. In practice, air at a low pressure is often referred to as a vacuum.

Valve A structure found in the heart and in blood vessels that prevents the back-flow of blood.

Variable resistor A circuit component whose resistance can be varied manually.

Variation Differences in the characteristics of organisms.

Vein A blood vessel that carries blood to the heart.

Velocity The speed and direction of a moving object.

Vena cava The main vein of the body, returning blood the the heart.

Ventricle A chamber in the heart that pumps blood out of the heart.

Vibration A to-and-fro or side-to-side movement such as that of a particle transmitting a wave.

Villus A finger-like projection in the small intestine which increases the surface area for the absorption of soluble food.

Virtual image An image that cannot be projected on a screen, as it does not exist.

Viscosity A measure of the ease with which a liquid can be poured. A liquid with a high viscosity, e.g. treacle, is difficult to pour.

Volt The unit of **voltage**. One volt is the voltage between two points if one joule of energy is transferred when one coulomb of charge moves from one point to the other.

Voltage This is needed to cause a current to pass in a conductor.

Voltmeter An instrument used to measure voltage.

W

Water A compound of hydrogen and oxygen with formula H_2O.

Watt The unit of power. One watt is equivalent to an energy transfer of one joule per second.

Wave A set of oscillations or vibrations that transfers energy without any transfer of mass.

Wavelength The distance occupied by one complete cycle of a wave.

Wax A shiny material that is impermeable to water.

Weathering The action of wind, rain, snow, etc. on rocks. These changes can be physical or chemical.

Weeds Unwanted plants growing in cultivated plots.

Weight The downward gravitational force acting on any object close to the surface of a planet. On Earth, each kilogram of mass has a weight of approximately ten newtons.

White cells Blood cells that protect us by engulfing microbes, producing antibodies and antitoxins.

White dwarf A phase in the life cycle of a small star during which helium nuclei fuse to form the nuclei of more massive elements.

Wilting The drooping of a plant through lack of water.

Work Is done when a force causes movement in its own direction. It always involves an energy transfer.

X

X chromosome A sex chromosome. Possession of two X chromosomes determines a female.

X-rays Very short wavelength electromagnetic radiation given off from X-ray machines.

Xylem A plant tissue largely responsible for the transport of water in the plant.

Y

Y chromosome A sex chromosome. Possession of one X and one Y chromosome determines a male.

Z

Zinc A dull, soft, silvery metal often used in alloys such as brass.

Zymase An enzyme in yeast which acts on sugar solution to produce ethanol by fermentation.

Answers

Life processes and blood

1. *Nutrition*; 2. *Respiration*; 3. *Excretion*; 4. *Reproduction*; 5. *Growth*; 6. *Sensitivity*; 7. *Movement*; 8. *Chromosomes*; 9. *Nucleus*; 10. *Cytoplasm*; 11. *Cell membrane*; 12. *Mitochondria*; 13. *Tissue*; 14. *Organ*; 15. *Organ system*; 16. *Pulmonary artery*; 17. *Pulmonary vein*; 18. *Aorta*; 19. *Vena cava*; 20. *Capillaries*; 21. *Haemoglobin*; 22. *Oxygen*; 23. *Carbon dioxide*; 24. *Soluble food*; 25. *Urea*; 26. *Hormones*; 27. *Ingest*; 28. *Antibodies*; 29. *Antitoxins*; 30. *Clot*.

Digestion

1. *Molar*; 2. *Premolar*; 3. *Canine*; 4. *Incisor*; 5. *Sharp*; 6. *Pointed*; 7. *Flat*; 8. *Oesophagus*; 9. *Liver*; 10. *Stomach*; 11. *Pancreas*; 12. *Small intestine*; 13. *Large intestine*; 14. *Gall bladder*; 15. *Insoluble*; 16. *Soluble*; 17. *Salivary glands*; 18. *Amylase*; 19. *Sugar*; 20. *Protease*; 21. *Amino acids*; 22. *Hydrochloric acid*; 23. *Lipase*; 24. *Fatty acids*; 25. *Insulin*; 26. *Bile*; 27. *Villi*; 28. *Surface area*; 29. *Glycogen*; 30. *Faeces*.

Control and co-ordination

1. *Eye*; 2. *Ear*; 3. *Skin*; 4. *Nose*; 5. *Ciliary muscles*; 6. *Cornea*; 7. *Pupil*; 8. *Lens*; 9. *Iris*; 10. *Optic nerve*; 11. *Retina*; 12. *Impulses*; 13. *Neurones*; 14. *Sensory*; 15. *Connector*; 16. *Effector*; 17. *Synapses*; 18. *Reflex action*; 19. *Effector*; 20. *Iris*; 21. *Cornea*; 22. *Ciliary muscles*; 23. *Hormones*; 24. *Endocrine*; 25. *Testosterone*; 26. *Oestrogen*; 27. *Insulin*; 28. *Glucose*; 29. *Diabetes*; 30. *High*.

Homeostasis

1. *Renal*; 2. *Kidney*; 3. *Artery*; 4. *Ureter*; 5. *Bladder*; 6. *Urethra*; 7. *Sweat*; 8. *Urine*; 9. *Ureter*; 10. *Bladder*; 11. *Urethra*; 12. *Skin*; 13. *Glomerulus*; 14. *Bowman's capsule*; 15. *Capillaries*; 16. *Collecting duct*; 17. *Filtration*; 18. *Glucose*; 19. *Water*; 20. *Urea*; 21. *ADH*; 22. *Increases*; 23. *Dialysis*; 24. *Transplant*; 25. *Dilate*; 26. *Constrict*; 27. *Evaporates*; 28. *Insulin*; 29. *Homeostasis*; 30. *Negative feedback*.

Reproduction, inheritance and health

1. *Nucleus*; 2. *Chromosomes*; 3. *Alleles*; 4. *In pairs*; 5. *Single*; 6. *X*; 7. *Y*; 8. *Girl−boy*; 9. *Dominant*; 10. *Recessive*; 11. *Homozygous*; 12. *Heterozygous*; 13. *Genotype*; 14. *TT*; 15. *Tt*; 16. *1 in 4*; 17. *Cc*; 18. *Nil*; 19. *Mitosis*; 20. *Meiosis*; 21. *Halves*; 22. *Meiosis*; 23. *Assortment*; 24. *DNA*; 25. *Bacteria*; 26. *Addiction*; 27. *Brain*; 28. *Cancer*; 29. *Cilia*; 30. *Depressant*.

Growth, variation and evolution

1. *Genes*; 2. *Environmental*; 3. *Genetic*; 4. *Continuous*; 5. *Discontinuous*; 6. *Asexual*; 7. *Identical*; 8. *Fertilisation*; 9. *Sexual*; 10. *Different*; 11. *Clones*; 12. *Chromosomes*; 13. *Alleles*; 14. *Mutations*; 15. *Radiation*; 16. *Cancer*; 17. *Selective breeding*; 18. *Characteristics*; 19. *Specialisation*; 20. *Mitosis*; 21. *The same number*; 22. *Breed*; 23. *Fertile*; 24. *Fossils*; 25. *Decay*; 26. *Evolution*; 27. *Variation*; 28. *Advantageous*; 29. *Darwin*; 30. *Lamarck*.

The environment

1. *Camouflage*; 2. *Insulation*; 3. *Thick*; 4. *Small*; 5. *Surface area*; 6. *Wax*; 7. *Nutrients*; 8. *Breeding*; 9. *Predators*; 10. *Prey*; 11. *Rise*; 12. *Fall*; 13. *Non-renewable*; 14. *Fossil fuels*; 15. *Combustion*; 16. *Pollution*; 17. *Carbon dioxide*; 18. *Sulphur dioxide*; 19. *Acidic*; 20. *Leaves*; 21. *Acidic*; 22. *Fertilisers*; 23. *Eutrophication*; 24. *Food chains*; 25. *Erosion*; 26. *Methane*; 27. *Carbon dioxide*; 28. *Radiation*; 29. *Ultraviolet*; 30. *CFCs*.

Particles and the periodic table

1. *Freezing*; 2. *Melting*; 3. *Evaporating*; 4. *Condensing*; 5. *Filtering*; 6. *Evaporating*; 7. *Distillation*; 8. *Fractional distillation*; 9. *Boiling points*; 10. *Chromatography*; 11. *Atoms*; 12. *Combined*; 13. *Compound*; 14. *Sulphide*; 15. *Nucleus*; 16. *Electrons*; 17. *Positively charged*; 18. *Protons*; 19. *Electrons*; 20. *Helium*; 21. *Inert*; 22. *Chlorine*; 23. *Period*; 24. *Group*; 25. *Atomic mass*; 26. *Atomic number*; 27. *Metals*; 28. *Non-metals*; 29. *Trends*; 30. *Alkali metals*; 31. *Halogens* 32. *Noble gases*; 33. *Hydrogen*; 34. *Alkali*; 35. *Iodine*; 36. *Chlorine*; 37. *Hydrogen chloride*; 38. *Chlorine*; 39. *Sodium hydroxide*; 40. *Shell*.

Reactivity series

1. *Reactivity*; 2. *Top*; 3. *Paraffin oil*; 4. *Oxygen*; 5. *Oxide*; 6. *Hydroxide*; 7. *Hydrogen*; 8. *Salt*; 9. *Hydrogen*; 10. *Copper*; 11. *Displacement*; 12. *Copper*; 13. *Iron*; 14. *Zinc*; 15. *Magnesium*; 16. *Corrosion*; 17. *Rust*; 18. *Iron(III) oxide*; 19. *Electrons*; 20. *Oxidation*; 21. *Reduction*; 22. *Zinc*; 23. *Copper*; 24. *Slowly*; 25. *Magnesium*; 26. *Quickly*; 27. *Electrolysis*; 28. *Reduction*; 29. *Uncombined*; 30. *Oxide*.

Acids, bases and salts

1. *Acids*; 2. *Sour*; 3. *Hydrogen*; 4. *Salt*; 5. *Acids*; 6. *Alkalis*; 7. *Neutral*; 8. *Universal*; 9. *Weak*; 10. *Strong*; 11. *Hydrogen*; 12. *Carbon dioxide*; 13. *Limewater*; 14. *Sulphate*; 15. *Salt*; 16. *Alkali*; 17. *Salt*; 18. *Nitric*; 19. *Sodium*; 20. *Sulphuric*; 21. *Sulphate*; 22. *Water*; 23. *Ammonia*; 24. *Carbon dioxide*; 25. *Sulphuric*; 26. *Sulphate*.

Rates of chemical reactions

1. *Very fast*; 2. *Explosion*; 3. *Very slow*; 4. *Decreases*; 5. *Collision*; 6. *Speed up*; 7. *Speeds up*; 8. *Collisions*; 9. *Decrease*; 10. *Increase*; 11. *Pressure*; 12. *Collisions*; 13. *Faster than*; 14. *Surface area*; 15. *Explosion*; 16. *Chlorine*; 17. *Catalyst*; 18. *Stays the same*; 19. *The same mass of product*; 20. *Finished*; 21. *Reactants*; 22. *Fastest*; 23. *Concentrated*; 24. *The same*; 25. *Less than*.

Chemicals from oil

1. *Organic*; 2. *Petroleum*; 3. *Gas*; 4. *Water*; 5. *Gas*; 6. *Impermeable rock*; 7. *Fossil*; 8. *Oxygen*; 9. *Carbon*; 10. *Detergents*; 11. *Fractional distillation*; 12. *Boiling points*; 13. *Crude oil vapour*; 14. *Refinery gases*; 15. *Petrol*; 16. *Paraffin*; 17. *Diesel oil*; 18. *Fuel oil*; 19. *Top*; 20. *Higher*; 21. *Alkanes*; 22. *Methane*; 23. *Covalent*; 24. *Saturated*; 25. *Increases*; 26. *Viscosity*; 27. *More difficult*; 28. *Carbon dioxide*; 29. *Carbon monoxide*; 30. *Cracking*; 31. *Catalyst*; 32. *Unsaturated*; 33. *Alkenes*; 34. *Ethene*; 35. *Colourless*; 36. *Ethene*; 37. *1,2-Dibromoethane*; 38. *Polymerisation*; 39. *Polymers*; 40. *Monomers*.

Using electricity

1. *Ammeter*; 2. *Energy*; 3. *Light*; 4. *Series*; 5. *Parallel*; 6. *Voltage*; 7. *Voltage*; 8. *Components*; 9. *Voltmeter*; 10. *Parallel*; 11. *Voltage*; 12. *Ohms*; 13. *Straight line*; 14. *Current*; 15. *Increases*; 16. *Decreases*; 17. *Alternating*; 18. *Live*; 19. *Neutral*; 20. *Earth*; 21. *Live*; 22. *Neutral*; 23. *Earth*; 24. *Earth*; 25. *Insulation*; 26. *Neutral*; 27. *Live*; 28. *Earth*; 29. *Neutral*; 30. *Earth*; 31. *Fuse*; 32. *Current*; 33. *Earth*; 34. *Resistance*; 35. *Plastic*; 36. *Conductors*; 37. *Insulated*; 38. *Earth*; 39. *Circuit breaker*; 40. *Fuses*; 41. *Energy*; 42. *Watts*; 43. *Voltage*; 44. *Power*; 45. *Kilowatt-hours*; 46. *Kilowatts*; 47. *Magnetic field*; 48. *Electromagnetic induction*; 49. *Magnet*; 50. *Electromagnet*.

Force and motion

1. *Speed*; 2. *Distance*; 3. *Gradient*; 4. *Speed*; 5. *Distance*; 6. *Gradient*; 7. *Speed*; 8. *Direction*; 9. *Directions*; 10. *Gravitational*; 11. *Weight*; 12. *Weight*; 13. *Air resistance*; 14. *Speed*; 15. *Weight*; 16. *Air resistance*; 17. *Terminal velocity*; 18. *Balanced*; 19. *Moment*; 20. *Pivot*; 21. *Force*; 22. *Force*; 23. *Elastic*; 24. *Springs*; 25. *Force*; 26. *Origin*; 27. *Proportional*; 28. *Hooke's*; 29. *Elastic limit*; 30. *Size*.

Electromagnetic radiation

1. *Cycle*; 2. *Wavelength*; 3. *Frequency*; 4. *Amplitude*; 5. *Hertz*; 6. *Electromagnetic spectrum*; 7. *Vacuum*; 8. *Frequency*; 9. *Radio waves*; 10. *Wavelength*; 11. *Frequency*; 12. *Gamma rays*; 13. *Frequency*; 14. *Wavelength*; 15. *Radio waves*; 16. *Microwaves*; 17. *Infrared*; 18. *Light*; 19. *Ultraviolet*; 20. *Gamma rays*; 21. *Gamma rays*; 22. *X-rays*; 23. *Wavelength*; 24. *Sound*; 25. *Reflection*; 26. *Wavelength*; 27. *Refraction*; 28. *Diffraction*; 29. *Wavelength*; 30. *Wavelengths*.

Energy resources and energy in reactions

1. *Fossil fuels*; 2. *Oil*; 3. *Non-renewable*; 4. *Energy*; 5. *Sun*; 6. *Temperature*; 7. *Turbines*; 8. *Generators*; 9. *Renewable*; 10. *Waves*; 11. *Turbine*; 12. *Hydroelectric*; 13. *Atmosphere*; 14. *Noise*; 15. *Renewable*; 16. *Earth*; 17. *Electricity*; 18. *Batteries*; 19. *Mains electricity*; 20. *Power*; 21. *Sun*; 22. *Efficiency*; 23. *Gravitational potential*; 24. *Gravitational potential*; 25. *Kinetic*; 26. *Exothermic*; 27. *Energy*; 28. *Oxygen*; 29. *Carbon*; 30. *Bonds*; 31. *Exothermic*; 32. *Endothermic*; 33. *Bond making*; 34. *Bond breaking*; 35. *Bond making*; 36. *Bond making*; 37. *Bond breaking*; 38. *Activation energy*; 39. *Catalyst*; 40. *Energy level diagrams*; 41. *Energy*; 42. *Reactants*; 43. *Activation energy*; 44. *Products*; 45. *Energy change*; 46. *Energy*; 47. *Reactants*; 48. *Activation energy*; 49. *Products*; 50. *Energy change*.

Transferring energy

1. *Heat*; 2. *Movement*; 3. *Sound*; 4. *Light*; 5. *Heat*; 6. *Electricity*; 7. *Efficiency*; 8. *Light*; 9. *Electricity*; 10. *Temperatures*; 11. *Radiation*; 12. *Evaporation*; 13. *Metals*; 14. *Insulators*; 15. *Gases*; 16. *Density*; 17. *Expands*; 18. *Electromagnetic radiation*; 19. *Infrared*; 20. *Emitters*; 21. *Absorbers*; 22. *Surface*; 23. *Energy*; 24. *Temperature*; 25. *Energy*; 26. *Conduction*; 27. *Convection*; 28. *Convection*; 29. *Conduction*; 30. *Reflects*.

Radioactivity

1. *Nucleus*; 2. *Alpha*; 3. *Beta*; 4. *Gamma*; 5. *Helium nucleus*; 6. *Geiger–Müller tube*; 7. *Electrons*; 8. *Aluminium*; 9. *Photographic film*; 10. *Electromagnetic radiation*; 11. *Lead*; 12. *Radiation*; 13. *Radioactive*; 14. *Radiation*; 15. *Background radiation*; 16. *Decreases*; 17. *Radiocarbon dating*; 18. *Radioactive*; 19. *Isotopes*; 20. *Beta*; 21. *Gamma*; 22. *Beta*; 23. *Gamma*; 24. *Random*; 25. *Nucleus*; 26. *Decay*; 27. *Half-life*; 28. *Half-life*; 29. *Quarter*; 30. *Background radiation*.

The Solar System

1. *Solar System*; 2. *Star*; 3. *Galaxy*; 4. *Stars*; 5. *Light*; 6. *Electromagnetic radiation*; 7. *Light-years*; 8. *Year*; 9. *Planets*; 10. *Sun*; 11. *Gravitational*; 12. *Orbital*; 13. *Year*; 14. *Axis*; 15. *Stars*; 16. *Seasons*; 17. *Mercury*; 18. *Pluto*; 19. *Sun*; 20. *Moon*; 21. *Satellites*; 22. *Orbit*; 23. *Orbit*; 24. *Geosynchronous*; 25. *Asteroid*; 26. *Jupiter*; 27. *Solar System*; 28. *Sun*; 29. *Comets*; 30. *Gravitational*.

Cell mechanisms and circulation

1. *Chloroplasts*; 2. *Cytoplasm*; 3. *Cell membrane*; 4. *Nucleus*; 5. *Cell wall*; 6. *Vacuole*; 7. *Cell sap*; 8. *Cellulose*; 9. *Chlorophyll*; 10. *Support*; 11. *Catalysts*; 12. *Optimum*; 13. *Carbon dioxide*; 14. *Ethanol*; 15. *Lactic acid*; 16. *Cell membrane*; 17. *Diffusion*; 18. *Concentration gradient*; 19. *Partially permeable*; 20. *Concentration gradient*; 21. *Osmosis*; 22. *Active transport*; 23. *Back-flow*; 24. *Muscular*; 25. *Elastic*; 26. *High*; 27. *Valves*; 28. *Plasma*; 29. *Oxygen*; 30. *Carbon dioxide*.

Breathing and respiration

1. *Trachea*; 2. *Rib*; 3. *Lung*; 4. *Diaphragm*; 5. *Alveoli*; 6. *Intercostal muscles*; 7. *Intercostal muscle*; 8. *Downwards*; 9. *Volume*; 10. *Pressure*; 11. *Oxygen*; 12. *Carbon dioxide*; 13. *Energy*; 14. *Aerobic*; 15. *Anaerobic*; 16. *Glucose*; 17. *Carbon dioxide*; 18. *Glucose*; 19. *Lactic acid*; 20. *Ethanol*; 21. *More*; 22. *Fatigue*; 23. *Oxidised*; 24. *Oxygen debt*; 25. *Lactic acid*; 26. *Water*; 27. *Mitochondria*; 28. *Moist*; 29. *Surface area*; 30. *Capillaries*.

Plant physiology

1. *Carbon dioxide*; 2. *Water*; 3. *Light*; 4. *Chlorophyll*; 5. *Chloroplasts*; 6. *Glucose*; 7. *Oxygen*; 8. *Respiration*; 9. *Energy*; 10. *Growth*; 11. *Cellulose*; 12. *Starch*; 13. *Lipids*; 14. *Amino acids*; 15. *Carbon dioxide*; 16. *Energy*; 17. *Oxygen*; 18. *Chlorophyll*; 19. *Surface area*; 20. *Chloroplasts*; 21. *Stomata*; 22. *Diffuse*; 23. *Xylem*; 24. *Phloem*; 25. *Carbon dioxide*; 26. *Temperature*; 27. *Auxins*; 28. *Elongate*; 29. *Weeds*; 30. *Cutting*; 31. *Roots*; 32. *Fruits*; 33. *Active transport*; 34. *Proteins*; 35. *Chlorophyll*; 36. *Root hair*; 37. *Xylem*; 38. *Transpiration*; 39. *Evaporates*; 40. *Stomata*; 41. *Guard cells*; 42. *Wilting*; 43. *Water*; 44. *Guard cells*; 45. *Osmosis*; 46. *Pressure*; 47. *Cell wall*; 48. *Turgor*; 49. *Out*; 50. *Plasmolysis*.

Energy flow and nutrient cycles

1. *Producers*; 2. *Herbivores*; 3. *Carnivores*; 4. *Omnivores*; 5. *Photosynthesis*; 6. *Consumers*; 7. *Energy*; 8. *A*; 9. *A*; 10. *B*; 11. *A*; 12. *C*; 13. *Fungi*; 14. *Recycling*; 15. *Photosynthesis*; 16. *Carbohydrate*; 17. *Respiration*; 18. *Respiration*; 19. *Microbes*; 20. *Respiration*; 21. *Combustion*; 22. *Faeces*; 23. *Movement*; 24. *Heat*; 25. *Putrefying bacteria*; 26. *Ammonium compounds*; 27. *Nitrifying bacteria*; 28. *Nitrates*; 29. *Nitrogen-fixing bacteria*; 30. *Denitrifying bacteria*.

Atoms and bonding

1. *Neutron*; 2. *Electron*; 3. *Proton*; 4. *Isotopes*; 5. *Isotopes*; 6. *Electrons*; 7. *Atomic number*; 8. *Mass number*; 9. *Sodium*; 10. *Chlorine*; 11. *Electrons*; 12. *Positively charged*; 13. *Molecular*; 14. *Giant structure*; 15. *Cubic*; 16. *Regular*; 17. *High*; 18. *High*; 19. *Low*; 20. *Low*; 21. *Bonding*; 22. *Sodium*; 23. *Chlorine*; 24. *Ions*; 25. *Lattice*; 26. *Electrostatic*; 27. *Molecule*; 28. *Pair*; 29. *Covalent*; 30. *Very weak*.

Quantitative chemistry

1. *Atom*; 2. *Hydrogen*; 3. *Mole*; 4. *Formula mass*; 5. *Atomic*; 6. *Molecules*; 7. *Volume*; 8. *Atmospheric*; 9. *Molar volume*; 10. *Concentration*.

Reversible reactions and transition metals

1. *Products*; 2. *Reversible*; 3. *Equilibrium*; 4. *Reverse reaction*; 5. *Ammonia*; 6. *Left*; 7. *Ammonia*; 8. *Right*; 9. *Reverse reaction*; 10. *Hydrogen*; 11. *Nitrogen*; 12. *Nitrogen*; 13. *Hydrogen*; 14. *Catalyst*; 15. *Iron*; 16. *Liquefying*; 17. *Recycled*; 18. *Exothermic*; 19. *Nitrogen*; 20. *Phosphorus*; 21. *Potassium*; 22. *Nitrogen*; 23. *Ammonia*; 24. *Nitric acid*; 25. *Sulphuric acid*; 26. *Quick acting*; 27. *Slow acting*; 28. *Photosynthesis*; 29. *Oxygen*; 30. *Eutrophication*; 31. *Transition metals*; 32. *Less reactive*; 33. *Denser*; 34. *High*; 35. *Ion*; 36. *Green*; 37. *Catalysts*; 38. *Copper*; 39. *Zinc*; 40. *Manganese*.

Electromagnetism

1. *Attract*; 2. *Repel*; 3. *Poles*; 4. *North*; 5. *Repel*; 6. *Attract*; 7. *Magnetic field*; 8. *North*; 9. *Electromagnets*; 10. *Current*; 11. *Soft iron*; 12. *Electromagnet*; 13. *Magnetic field*; 14. *Armature*; 15. *Contact screw*; 16. *Armature*; 17. *Current*; 18. *Armature*; 19. *Attracted*; 20. *Magnetic fields*; 21. *Alternating*; 22. *Current*; 23. *Voltage*; 24. *Transformer*; 25. *Primary*; 26. *Secondary*; 27. *Secondary*; 28. *Primary*; 29. *Magnetic field*; 30. *Current*.

Electrostatics and current

1. *Electrons*; 2. *Forces*; 3. *Repel*; 4. *Attract*; 5. *Friction*; 6. *Electrons*; 7. *Negatively*; 8. *Positively*; 9. *Conductors*; 10. *Static*; 11. *Charge*; 12. *Voltage*; 13. *Sparks*; 14. *Conductor*; 15. *Electrostatic*; 16. *Current*; 17. *Electrons*; 18. *Negative*; 19. *Positive*; 20. *Negative*; 21. *Energy*; 22. *Power*; 23. *Watts*; 24. *Voltage*; 25. *Coulomb*; 26. *Time*; 27. *Charge*; 28. *Charge*; 29. *Coulomb*; 30. *Volt*.

Force and energy

1. *Work*; 2. *Force*; 3. *Distance*; 4. *Joules*; 5. *Conserved*; 6. *Watt*; 7. *Power*; 8. *Speed*; 9. *Balanced*; 10. *Force*; 11. *Equal*; 12. *Opposite*; 13. *Friction*; 14. *Resistive*; 15. *Braking distance*; 16. *Speed*; 17. *Mass*; 18. *Stopping distance*; 19. *Thinking distance*; 20. *Reaction time*; 21. *Mass*; 22. *Balanced*; 23. *Driving*; 24. *Resistive*; 25. *Unbalanced*; 26. *Accelerating*; 27. *Velocity*; 28. *Mass*; 29. *Height*; 30. *Mass*.

Pressure and sound

1. *Pressure*; 2. *Force*; 3. *Small*; 4. *Pressure*; 5. *Large*; 6. *Force*; 7. *Pressure*; 8. *Forces*; 9. *Hydraulic*; 10. *Forces*; 11. *Pressure*; 12. *Proportional*; 13. *Random*; 14. *Pressure*; 15. *Halves*; 16. *Proportional*; 17. *Longitudinal*; 18. *Vibrating*; 19. *Frequency*; 20. *Amplitude*; 21. *Light*; 22. *Vacuum*; 23. *Echoes*; 24. *Ultrasound*; 25. *Echo*; 26. *Speed*; 27. *Ultrasound*; 28. *Reflections*; 29. *Speed*; 30. *Diffracted*.

The Earth, its atmosphere and waves

1. *Mixture*; 2. *Oxygen*; 3. *Respire*; 4. *Burned*; 5. *Photosynthesis*; 6. *Increased*; 7. *Greenhouse effect*; 8. *Global warming*; 9. *Galaxies*; 10. *Big Bang*; 11. *Universe*; 12. *Energy*; 13. *Longitudinal*; 14. *Vibrations*; 15. *Sound*; 16. *Longitudinal*; 17. *Light*; 18. *Transverse*; 19. *Frequency*; 20. *Critical angle*; 21. *Total internal reflection*; 22. *Wavelength*; 23. *Transverse*; 24. *P*; 25. *S*; 26. *Seismometers*; 27. *P*; 28. *Core*; 29. *Core*; 30. *P*.

Rocks in the Earth

1. *Igneous*; 2. *Crystals*; 3. *Slowly*; 4. *Intrusive*; 5. *Crystals*; 6. *Quickly*; 7. *Extrusive*; 8. *Sedimentary*; 9. *Older*; 10. *High*; 11. *High*; 12. *Metamorphic*; 13. *Marble*; 14. *Calcium carbonate*; 15. *Igneous*; 16. *Igneous*; 17. *Magma*; 18. *Metamorphic*; 19. *Sedimentary*; 20. *Intrusion*; 21. *Igneous*; 22. *Sedimentary*; 23. *Metamorphic*; 24. *Fossils*; 25. *Radioactivity*; 26. *Crust*; 27. *Mantle*; 28. *Convection*; 29. *Core*; 30. *Plates*; 31. *Slide*; 32. *Faults*; 33. *Boundaries*; 34. *Constructive*; 35. *Magma*; 36. *Destructive*; 37. *Oceanic*; 38. *Continental*; 39. *Magma*; 40. *Subduction*.

Using rocks and ores

1. *Open cast*; 2. *Rock*; 3. *Haematite*; 4. *Bauxite*; 5. *Electrolysis*; 6. *Reduction*; 7. *Uncombined*; 8. *Blast*; 9. *Coke*; 10. *Air*; 11. *Carbon*; 12. *Carbon dioxide*; 13. *Reduced*; 14. *Reducing agent*; 15. *Electrolysis*; 16. *Cryolite*; 17. *Carbon*; 18. *Cathode*; 19. *Aluminium*; 20. *Anode*; 21. *Oxygen*; 22. *Burn*; 23. *Carbon dioxide*; 24. *Electrolysis*; 25. *Anode*; 26. *Cathode*; 27. *Electrolyte*; 28. *Cathode*; 29. *Anode*; 30. *Ions*; 31. *Steel*; 32. *Electricity*; 33. *Alloy*; 34. *Electricity*; 35. *Expensive*.